Erik Seedhouse

Tourists in Space

A Practical Guide

 Springer

Published in association with
Praxis Publishing
Chichester, UK

Dr Erik Seedhouse FBIS, AsMA
Sooke
British Columbia
Canada

SPRINGER–PRAXIS BOOKS IN SPACE EXPLORATION
SUBJECT *ADVISORY EDITOR*: John Mason, M.Sc., B.Sc., Ph.D.

ISBN 978-0-387-74643-2 Springer Berlin Heidelberg New York

Springer is part of Springer-Science + Business Media (springer.com)

Library of Congress Control Number: 2007939915

Cover design: Jim Wilkie
Project management: Originator Publishing Services Ltd, Gt Yarmouth, Norfolk, UK

Printed on acid-free paper

Tourists in Space
A Practical Guide

Contents

TMO	Trash Management Operations
TPR	Total Peripheral Resistance
TPS	Thermal Protection System
TS	Trauma Sonography
TTD	Tilt Translational Device
TTT	Tilt Table Test
TUC	Time of Useful Consciousness
TV	Tidal Ventilation
TV	Tidal Volume
TVC	Thrust Vector Control
TVIS	Treadmill Vibration Isolation and Stabilization System
UCD	Urine Collection Device
UV	Ultraviolet
VAFB	Vandenberg Air Force Base
VAPAK	Vapor pressurization design
VCG	Vectorcardiograph
VE	Virtual Environment
VEG	Virtual Environment Generator
VF	Ventricular Fibrillation
VLA	Very Large Aircraft
VR	Virtual Reality
VRI	Visual Reorientation Illusions
VSS	Virgin SpaceShip
VT	Ventricular Tachycardia
VTO	Vertical Take Off
VTOVL	Vertical Take-off Vertical Landing
WCS	Waste Collection System
WK1	WhiteKnightOne
WK2	WhiteKnightTwo
WMC	Waste Management Compartment
WMS	Waste Management System
WS	Waist Straddle

Introduction

Commercial potential for space tourism

Man must rise above Earth to the top of the atmosphere and beyond, for only then will he fully understand the world in which he lives.

Socrates, sometime around 400 BC

DEFINING SPACE AND SPACE TOURISTS

Before addressing the topic of this chapter, it is important to agree on the definition of a space tourist and what exactly constitutes space. The description *astronaut, cosmonaut, spationaut,* and *taikonaut* are terms generally reserved for professional space travelers, trained by a human spaceflight program provided by a government space agency such as the Russian Federal Space Agency, NASA, or ESA, to serve as a crewmember of a spacecraft. Until the birth of the orbital space business in 2001, with the flight of Dennis Tito, professional space travelers were trained exclusively by government space agencies, but Tito's flight created a new category of space traveler. Following his pioneering flight, Tito was described by the press as a commercial astronaut, a space tourist, a civilian astronaut, a pseudo-astronaut, a private space explorer, and a *spaceflight participant,* the latter of which is probably the most appropriate definition for those paying to travel into space. It is also the term agreed on by NASA and the Russian Federal Space Agency to distinguish space travelers from professional astronauts. Since *spaceflight participant* is also the designation adopted by the Federal Aviation Administration (FAA) in its regulations governing commercial spaceflight, it is this term that will be used in this guide when referring to those paying for their trip into space.

Another important definition is what constitutes space. According to the Fédération Aéronautique Internationale (FAI) [5], space begins at an altitude of 100 km, whereas the United States Air Force defines space as any altitude over

80 km and even awards astronaut wings to personnel reaching this height. Since the Ansari Prize required an operator to achieve an altitude of 100 km and commercial space companies have adopted the FAI guideline, it is this definition that will be used in this book. At the time of writing, using the FAI definition, 460 people from 39 countries have traveled into space, of whom 456 have reached Earth orbit, and 24 people have traveled beyond Low Earth Orbit (LEO). As we shall see in this chapter, based on current market research, these numbers are likely to increase dramatically in the very near future.

THE DEMAND FOR SPACE TOURISM

For several years, interest in the potential of space tourism has been steadily increasing among engineers, scientists, entrepreneurs, and the general public, as evidenced by the growing number of articles and publications addressing the subjects of reusable launch vehicles (RLVs), space habitats, and spaceports. The attractiveness and exclusivity of traveling into space are two powerful driving forces that may, in the near future, turn space tourism into the multi-billion dollar business predicted by several market research surveys.

Since much of the content of this guide is driven largely by the demands and perceptions of future spaceflight participants, it is appropriate to review current market research that has attempted to define the requirements of this emerging industry.

The Futron/Zogby Poll

Demand for passenger space travel has been determined by a study conducted by Zogby International for the Futron Corporation, a Wisconsin-based company that published the *Space Tourism Market Study* report in 2006 [12], the results of which were recently updated to take into account the success of SS1 (Figure I.1).

Figure I.1 predicts the passenger demand for suborbital flights, based on factors such as projected ticket prices, space qualification requirements, and training time. Based on studies of the public's willingness to pay, analysts predict the space suborbital tourism market will produce revenues in excess of U.S.$700 million[1] by 2021 [1].

> "Ultimately, a passenger's ticket price for a suborbital spree will come down after several hundred people have flown ... in 1990, a T-1 connection to the Internet cost $1 million a year. Now the equivalent service is $19.99 a month."
>
> Rich Pournelle, Director of Business Development, XCOR Aerospace
> in interview with *Ad Astra*, June 6th, 2005

The Adventurers' Survey

Another notable study of public perceptions was an online survey conducted in 2006 by Derek Webber of Spaceport Associates and Jane Reifert of Incredible Adventures,

---

[1] All dollar amounts are U.S. dollars.

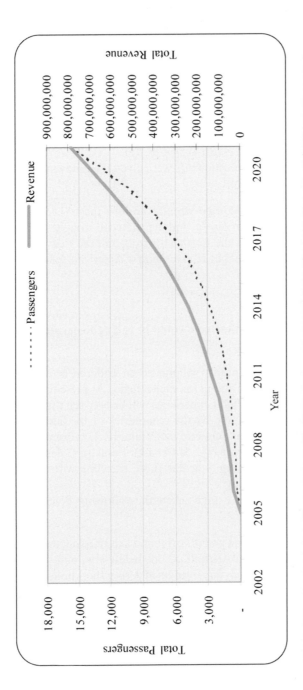

	2006	2007	2008	2009	2010	2011	2012	2013	2014	2015	2016	2017	2018	2019	2020	2021
Passengers	356	455	591	769	999	1,298	1,685	2,186	2,830	3,656	4,711	6,048	7,770	9,916	12,545	15,712
Price (US$ K)	100	100	100	100	100	100	95	90	85	80	75	70	65	60	55	50
Rev. (US$ M)	36	46	59	77	100	130	160	197	241	293	353	423	505	595	690	786

Figure I.1. Passenger interest and demand forecast for suborbital space tourism. Image courtesy: Futron Corporation. Source: *http://www.futron.com/pdf/resource_center/conference_presentations/Starzyk%20-%20RAeS.pdf*

which obtained and analyzed 998 responses. Whereas the Futron survey was conducted in a statistically valid and unbiased way in order to make it possible to derive forecasts of demand, the survey conducted by Webber and Reifert focused on people biased in favor of adventure holidays, which is why the survey is often referred to as *The Adventurers' Survey* [8, 15]. Although *The Adventurers' Survey* cannot forecast demand, the opinions of the respondents are relevant to those operators engaged in designing the space tourism experience.

The survey investigated three aspects of space tourism in which client perception had the potential to provide input for space tourism operators. The aspects were *training*, *orbital duration*, and *spacecraft design* [8, 15].

For the training component, respondents were given the following description of what constitutes a suborbital flight:

> "This is an adventure where you go straight up into space (i.e. above 60 miles high), spend about 5 minutes in zero-g, and then descend straight back again. You see black sky, the curvature of the Earth's horizon, and the view for hundreds of miles in every direction. This flight is like the first American spaceflight by Alan Shepard." [15]

Respondents were then asked:

> "What is the most amount of time you would be willing to devote to a suborbital space adventure?" [15]

The response indicated a strong public willingness to undergo training, with 59% willing to spend up to two weeks training for such a flight. At present, the anticipated training duration for a suborbital flight experience will be between three days and two weeks, a period that is driven largely by the complexity of the launch vehicle and whether there is a requirement for pressure suit training. For example, the British and U.S.-based space tourism company, Starchaser Industries, demand that their passengers spend two weeks training, whereas those traveling with Virgin Galactic will require only three days training.

Orbital flight respondents were then given the following description of what constitutes such a flight:

> "This is an adventure where you go into orbit and keep circling the Earth every 90 minutes. You will see all countries and oceans below you, and will experience a sunrise and sunset 16 times each day, and be in zero-g the whole time. This flight gives you the same views and experiences as today's Shuttle astronauts. In some of the flights, there will be docking with a space hotel. In others, you will remain in the spaceship that brought you." [15]

Respondents were then asked:

> "What's the most amount of time you'd be willing to devote to preparing for and completing an Orbital Space Mission, including medical checks, training and the flight itself." [15]

Thirty percent of respondents indicated a maximum acceptable duration of 3 months for training, a period that is less than half the current period of preparation required by civilian astronauts *en route* to the International Space Station (ISS). Clearly, for orbital flight to be attractive to customers, the length of training will need to be reduced to a more manageable timeframe of between four to six weeks.

To determine the preferred length of flight, the survey asked respondents how long they wanted to spend in orbit, a question to which 70% of respondents indicated they would be happy with two weeks or less.

The survey also investigated the importance of and location of spaceports in the context of suborbital space tourism by asking the following question:

"Spaceports—How much do they matter? Does the location of a suborbital spaceport matter? Spaceports are the locations where spaceflights will begin and end." [15]

Forty-eight percent of respondents indicated that they would be prepared to go anywhere, while another 31% indicated they would also go anywhere, provided the spaceport was located in their country.

For those passengers embarking on an orbital flight, the destination will be an important consideration as the choice will be either to rendezvous with a space hotel/station or to remain in the spacecraft. This consideration was investigated by asking the following question:

"How important is a stop at a space hotel or space station? In other words, is a destination in space important or are you happy staying in a spacecraft the whole mission?"

The unpredictable response to this question was that 79% indicated they were not interested in a hotel option and of the 21% who indicated that a hotel was important, 73% indicated a willingness to pay a premium of up to 20% on the original price [15].

Another factor addressed in the survey was an assessment of the financial willingness to pay. Whereas only 7% of the respondents indicated they would be interested in undertaking the flights at the $190,000 ticket price currently advertised by Virgin Galactic, 36% indicated they would be interested if the price was reduced to below $50,000. A similar assessment was made with respect to orbital flight, in which 14% indicated they considered $5 million to be a fair price for such a trip.

In assessing when respondents would want to fly, the survey asked whether people would be prepared to pay extra for the privilege of being one of the first to fly, if they would wait for the price to come down before flying, or to wait until the technology is proven. Based on these questions, 47% of respondents indicated they would prefer to wait until the price had dropped [15].

Clearly, based on the responses to *The Adventurers' Survey* [8] and on the information contained in the *Space Tourism Market Study*, a significant demand exists for both suborbital and orbital space tourism. It is also clear, based on the performance of SS1, that there exists the potential to satisfy this demand, but for future personal spaceflight companies/commercial launch operators to succeed there are several challenges that must be addressed, such as safety and the myriad legal and bureaucratic hurdles that form part of the regulatory environment.

REGULATORY ISSUES

Perhaps the most challenging issue faced by commercial launch operators is the regulatory environment, with its potential to discourage private sector investment and its concomitant detrimental effect. Some examples of the regulatory issues faced by a commercial launch operator include

1.   Launch site, environmental, and range safety approvals for launch and recovery.
2.   International Trade on Arms Regulations (ITAR) export and control.
3.   FAA licensing for new launch vehicles.

The above issues represent significant challenges to the nascent space tourism industry, and while a comprehensive discussion of these matters is beyond the scope of this manual, it is appropriate to provide a brief overview of the means that exist for commercial launch operators to overcome such obstacles.

For example, with the publication of the Aldridge Commission report [11], commercial launch operators now have assistance in overcoming some of the hurdles mentioned. The Aldridge Commission was created to examine and make recommendations on implementing a new vision for space exploration with particular attention being paid to the role of private industry, as stated in one of the recommendations of the report:

> "The Commission recommends NASA recognize and implement a far larger presence of private industry in space operations with the specific goal of allowing private industry to assume the primary role of providing services to NASA, and most immediately in accessing low-Earth orbit."
>
> Aldridge Commission, Recommendation 3-1, p. 19

Furthermore, the Aldridge Commission makes recommendations that encourage NASA to create incentives for entrepreneurial investment in space so that commercial space companies do not have to shoulder the whole cost of a launch vehicle. It is anticipated that implementation of the Commission's recommendations will accelerate the development of technologies and systems that enable suborbital and orbital access. In so doing it will lay the foundations to create a truly independent space industry, instead of the various government-funded space programs and their vendors. However, although the recommendations of the Commission will assist commercial space companies overcoming regulatory issues, other hurdles exist in the form of requirements imposed by the FAA and the absence of legal guidance in the areas of insurance and liability.

LEGAL CONSIDERATIONS AND REGULATIONS

FAA requirements

In December 2006, the FAA established requirements for human spaceflight as required by the Commercial Space Launch Amendments Act (CSLAA) of 2004 [2, 4]. The regulations, entitled "Human Space Flight Requirements for Crew and Space Flight Participants", came into effect February 13th, 2007, and include rules concerning crew qualifications and training, and informed consent for crew and spaceflight participants. The rules are based partly on the comments the FAA received from 42 entities in response to its Notices of Proposed Rulemaking (NPRM) on December 23rd, 2005. The comments included those made by aerospace companies, service providers, and operators of launch and re-entry vehicles such as Blue Origin, TGV Rockets, and XCOR Aerospace. Other associations and service providers who commented included the Airline Pilots Association International (ALPA), Association of Space Explorers-U.S.A. (ASE), International Association of Space Entrepreneurs (IASE), and the Institute for Space Law and Policy (ISLAP). The requirements outlined in the CSLAA assist operators in determining what they must and must not do in terms of training their passengers and crew and also what medical certification is required, but it does not provide any guidance on the issues of insurance and liability.

Liability in space

Several laws already exist to regulate private sector space endeavors, the most significant being the Outer Space Treaty (OST) of 1967 [14] which, although stipulating the principle of "exploration and the use of outer space", does not provide operators with much legal guidance. The treaty considers astronauts as "envoys of humankind" and their status as astronauts is clearly defined as being associated with the nation to which they are registered. However, this definition does not apply to spaceflight participants, who do not represent any nation and do not conduct any activity for any national entity, although spaceflight participants Dennis Tito, Greg Olsen, Anousheh Ansari, and Mark Shuttleworth all conducted experiments onboard ISS. Since there is no legal definition of a spaceflight participant's status there is also an absence of rules that regulate rights and duties, causing problems when it comes to defining responsibility and liability. For example, the liability of the launching state for damages caused by a space object is not clearly stated, so it is uncertain which liability regime would be applied in the event that a non-governmental entity's spacecraft were to cause damage to a foreign citizen or foreign property!

Another problem faced by space lawyers and the operators that will employ them is the question of liability in a case in which a spaceflight participant engages in extravehicular activity (EVA). In such a case, lawyers will need to identify who holds jurisdiction over the spaceflight participant while outside the vehicle. This may prove difficult, since, under the terms of the OST, a spaceflight participant engaging in EVA outside a vehicle is considered a space object (!) and under the OST regulations, jurisdiction does not extend outside the physical limits of the structure of a vehicle! The Russian Federation on Space Activity solved this quandary by stipulating that the Russian Federation would "retain jurisdiction and control over a crew of a manned space object registered in it, during the ground time of such object, at any stage of a spaceflight or stay in space, on celestial bodies, *including extra-vehicular stay*, and return to Earth, right up to the completion of the flight program, unless otherwise specified in the international treaties of the Russian Federation." This regulation is fine if you happen to be a fare-paying passenger onboard a Russian vehicle, but it does not help if you are a "space object" outside another state's spacecraft! The problem faced by space lawyers is that traditionally, the launching state retains jurisdiction over personnel onboard a spacecraft but spaceflight parti-cipants are not considered personnel, but passengers [1]. To resolve this problem it is possible the space tourism industry will adopt the 1929 Warsaw Convention [3] to define the legal status of a spaceflight participant, since this convention sets the operator's responsibility above that of its passengers. Clearly, in the next few years there needs to be a common legal regime governing a crew code of conduct for spaceflight participants so that a hierarchical relationship can be established in orbit and with those on the ground. Specific "rules of the road" will need to be established, so that the spaceflight participant is submitted to the jurisdiction and control of the state of registration of the spacecraft on which he/she is onboard and also to the commander's authority when applicable. In light of the above, for those purchasing a ticket in the next three to five years, it would behoove you to check the legal aspects pertaining to liability.

Licensing

Other legal considerations that must be addressed are those related to licensing. At the time of writing, there is a large gap between rocket and aircraft reliability, a gap that will remain for some time, due to the complex nature of flying into space. Whereas it is possible to calculate with a high degree of accuracy the airworthiness of an aircraft type, the reliability of a spacecraft can only be estimated by a reliability calculation, which is how the Space Shuttle's reliability is calculated. Clearly, the certification process that is applied to civil aviation, in which more than 1,000 test flights are typically required to gather sufficient data for certification, cannot be applied to space transportation systems. The probability of loss of a spacecraft is therefore a function of the failure rate that will result in either catastrophic loss of the vehicle or a failure of one or more subsystems within the vehicle. Clearly, it will not be possible to permit certification of space transportation vehicles that carry the same prediction of failure as the Space Shuttle, which means that the existing regulations need to be reformed, which in turn will probably result in a restriction in the design of some space vehicles.

The change in regulations has implications not only for attracting customers but also for investors, since any investor will want to understand and control their capital risk, which is only possible in a regulatory market environment. Licensing space transportation systems will also require certification regulations for structural integrity, fire suppression systems, environmental control systems, emergency egress standards, and maintenance standards and procedures. It is possible that eventually, the civil aviation model will be adapted to fulfill these requirements, but even if the licensing problem is resolved the problem of third-party and other insurance issues remains.

Insurance

Under the CSLAA the launch vehicle prime contractor is obliged to purchase insurance to cover maximum probable loss (MPL) for injury or damage to third parties. This insurance may cost the contractor up to U.S.$500 million for third-party injury or damage and up to U.S.$100 million for damage to government property. The MPL for manned suborbital flights will probably be in the region of U.S.$10 million [10], which will severely reduce the operators' ability to purchase insurance, as the revenues of a typical suborbital flight will be only a hundredth of a conventional commercial flight. There have already been examples of the severity of MPL coverage in the nascent commercial launch industry, one of which involved the Rotary Rocket Company, which had to pay U.S.$150,000 for U.S.$1 million of MPL coverage for its Roton experimental vehicle. Another commercial launch company, Armadillo Aerospace, had been asked to pay underwriters U.S.$1 million for a U.S.$9 million MPL when testing its experimental vehicle, although they eventually obtained the coverage for less. The most prominent private space company, Scaled Composites, paid only U.S.$5,000 for U.S.$1 million of MPL coverage for SS1, although this premium will most likely be closer to U.S.$500,000 for the MPL coverage of SpaceShipTwo (SS2).

On paper, the FAA rules, which were designed with the intent of promoting commercial space travel, may militate against any problems associated with high premiums, but the reality may be different. The rules require that each crewmember and passenger be informed in writing of the risks involved and to waive their rights to claim for damages, provided operators show "acceptable levels of safety". However, given the net worth of the individuals who will be flying on suborbital and orbital flights it is likely that in the event of a catastrophic failure of launch vehicle, family members will sue anyway.

SAFETY

"The pursuit of discovery is a risky business, and it will continue to be so for the foreseeable future. Astronauts know this and accept the risk."

Aldridge Commission

One question that is probably on the minds of many people is: "How safe will the personal spaceflight industry be?" Obviously, the private space companies know they cannot afford to have a significant failure, since loss of life would be devastating to not only their company but also to the industry. As with any industry, safety legislation provides guidelines and regulations, which, for the private spaceflight industry, is contained in the CSLAA [2, 4]. At a national level, there is the National Space Transportation Policy that gives the FAA authority to issue licences and experimental permits for testing and development. However, despite all the regulations and legislation and despite all the safety procedures, the fact remains that the first suborbital passengers will be embarking on a journey that is several orders of magnitude more dangerous than an airline flight. Even today, there are only 4 to 6 manned spaceflights (combined Space Shuttle and Soyuz flights) and between 10 and 15 commercial satellite launches per year, and there are still launch failures despite extensive ground and support facilities. Even NASA, which is often criticized by space entrepreneurs for emphasizing safety over innovation, has a fatality rate of 6% [13] for manned spaceflight, compared with an equivalent fatality rate of 0.0000002% for commercial aviation. That risk, which has been present during the entire Space Age, may even grow as the industry is privatized, since private individuals may be willing to take risks that government agencies can't take. It should also be remembered that for government-funded space agencies, spaceflight has always been a matter of national prestige which has meant that NASA has traditionally been reluctant to stretch the risks of both their astronauts and their spacecraft. Yet, even with this culture of safety, accidents have still occurred. The images of the Space Shuttle Challenger exploding in 1986 and the tragic images of Columbia's disintegration in 2003 inevitably fuel the question, "Aren't people going to die?" and a comparison of the personal spaceflight industry with the aviation industry 100 years ago.

In the early days of air travel, aviation barnstormed technology forward despite fatalities and crashed planes in an environment that was short on regulation and light on legal liability. Unlike the aviation industry, however, the personal spaceflight equivalent is being born in a very different regulatory and legal environment that makes it much more difficult for a company to implement any procedure or machine that may compromise safety. Also, the personal spaceflight industry has the benefit of being able to draw on the legacy of aviation and learn from the mistakes that were made, a situation that is obviously of benefit to future spaceflight participants. Nevertheless, despite operating in a stricter regulatory environment, private space companies will need to be cognizant of legal liability, government regulation, and concentrate on safety with a persistent and laser-like focus. Even Burt Rutan, the maverick engineering genius behind SS1, admits there is no way that SS2 will ever be as safe as a 747 airliner. He and his team at Scaled Composites are realistically aiming at achieving the safety rate of the early airliners, recognizing that space travel is a risky enterprise and that sometime during the next five or ten years there will inevitably be an event in which a loss of vehicle occurs.

"We must live with these risks and the possibility to fail. Without taking risks, there are no breakthroughs."

Peter Diamandis

At this point it is appropriate to put the risk of space travel in the perspective of other adventurous activities. Mountaineers, for example, who attempt to climb Mt. Everest, have a 1 in 12 chance of dying [6], whereas those who attempt K2 historically have a 1 in 3 chance of not returning [7]. These risk levels compare favorably with the present 6% chance of dying in human spaceflight, although current private space companies will need to improve on this if they want to attract spaceflight participants in large numbers! Despite these risks thousands of people attempt Mt. Everest every year and every time there is an astronaut selection NASA receives more than 5,000 applications. There is clearly an unmet demand to go into space.

People around the world are talking about space transportation and quickly beginning to realize that commercial space transportation, as an industry, is fast becoming a reality. Hopefully the missions of Virgin Galactic will collectively and iteratively build the body of knowledge required to ensure the industry is a safe one, although recent events such as the death of three Virgin Galactic workers in July 2007 has cast a cloud over the industry's safety record. With time what may seem impossible today will soon become commonplace and what is intrinsically governmental today will eventually become independent. Eventually, spurred in large part by the endeavors of "entreprenaut" companies such as Virgin Galactic, XCOR Aerospace, and Rocketplane Ltd., commercial spaceflight will be as common as air travel and suborbital and orbital space tourism will emerge as a truly successful market. Ultimately, the private space industry will create a reliable orbital infrastructure that will serve as an industrial base and a gateway to space. For those interested in traveling there, this manual will explain what you must do.

REFERENCES

[1] Bhattacharya, J. Legal aspects of Space Tourism. *Proc. of the 45th Colloq., Houston, 2002,* p. 212.

[2] Commercialization of Space Commercial Space Launch Amendments Act of 2004. H.R. 3752, 108th Cong.

[3] *Convention for the Unification of Certain Rules relating to International Carriage by Air, Warsaw, October 12th, 1929.*

[4] Department of Transportation. Federal Aviation Administration. 14 CFR Parts 401, 415, 431, 435, 440, and 460. Human Space Flight Requirements for Crew and Space Flight Participants: Final Rule. FAA. December 15, 2006.

[5] *www.fai.org/astronautics/*—Fédération Aéronautique Internationale. FAI Sporting Code, General section.

[6] Huey, R.B.; and Salisbury, R. Success and Death on Mt Everest. *American Alpine Journal,* **287**, 1–10 (2003).

[7] Huey, R.B.; and Eguskitza, X. Supplemental Oxygen and Death rates on Everest and K2. *Journal of the American Medical Association*, **284**, 181 (2000).

[8] *www.incredible-adventures.com/space-survey/space-adventurers-survey.pdf*

[9] *www.nasa.gov/pdf/60736main_M2M_report_small.pdf*—Aldridge.

[10] Pagnanelli, B. *14th International Space Insurance Conference, Milan, March 22–23, 2007.*

[11] Report of the President's Commission on Implementation of United States Space Exploration Policy (June 2004).

[12] *www.spacetourismsociety.org/Presentations_files/SpaceTourismMarketStudy.pdf*

[13] *www.thespacereview.com/article/36/2*

[14] Treaty on Principles Governing the Activities of States in the Exploration and Use of Outer Space including the Moon and other Celestial Bodies. Washington, London, Moscow (January 27, 1967).

[15] Webber. D.; and Reifert. J. *Executive Summary of the Adventurers' Survey of Public Space Travel.* Spaceport Associates, MD (September 2006).

1

Strapping rockets to dreams: The significance of SpaceShipOne

We want our children to go to the planets. We are willing to seek breakthroughs by taking risks.

Burt Rutan, shortly before SpaceShipOne's historic flight

THE FLIGHT OF N328KF

October 4th, 2004. A historic event is taking place at Mojave Airport, a sprawling civilian test center in the California high desert 150 kilometers from Los Angeles, where hundreds of rusting aircraft, their engines and undercarriages shrink-wrapped, sit parked in long, lonely rows. But on this Monday morning the motley collection of DC10s, 747s, DC9s, and 737s, representing airlines from around the world, will bear witness to a truly extraordinary event. Here, at this desolate airport, a small, winged spacecraft built with lightweight composites and powered by a rocket motor using laughing gas and rubber will fly to the edge of space and into the history books. Registered with the Federal Aviation Administration (FAA) only by the anonymous designation N328KF1,[1] but known to space enthusiasts as SpaceShipOne (SS1) and its carrier vehicle, WhiteKnight, this privately developed manned vehicle (Figure 1.1) will finally open the door for a much greater portion of humanity waiting to cross the threshold into space.

The excitement began building the night before, as cars poured into the parking lot and continued to stream in almost until takeoff, by which time crowd control personnel had almost been overwhelmed. Rows of trucks with satellite dishes and

[1] The "N" in the designation is the prefix used by the FAA for U.S.-registered aircraft and the 328KF stands for 328 (kilo, "K") thousand feet (the "F" in the designation), which is the official demarcation altitude for space.

Figure 1.1. SpaceShipOne and WhiteKnightOne. Image courtesy: © 2004 Mojave Aerospace Ventures LLC, photograph by Scaled Composites. SpaceShipOne is a Paul G. Allen Project. Source: *www.scaled.com/projects/tierone/gallery/X-Prize_1/XPrize_X1_0166*

glaring spotlights greet the spectators as they stream into the airport. It is only five in the morning but a sense of expectancy already wafts through the air together with the smell of coffee and bagels. A huge X-Prize banner flutters from the control tower, as thousands of space enthusiasts from around the world wait for the Sun and the appearance of WhiteKnight. Legends of the space program, such as Buzz Aldrin, mill around in the VIP area together with William Shatner and Mojave's resident maverick engineering genius, Burt Rutan. Only a few miles away at Edwards Air Force Base on August 22nd, 1963, test pilot Joe Walker reached the edge of space by flying an Air Force X-15 rocket plane to an altitude of 107,333 meters. The X-15 eventually gave birth to the Space Shuttle, a semi-reusable vehicle embroiled in politics that became a symbol that the high frontier was the absolute dominion of governments and space agencies, a *status quo* perpetuated for more than three decades. Until now. More than 40 years after Walker's flight, using a flight profile similar to the X-15's, SS1 will attempt to beat Walker's record. Today, on the 47th anniversary of Sputnik, a privately developed spacecraft will attempt to demonstrate that it is not necessary to spend U.S.$20,000 to put one kilogram into orbit, or to have the technologies of space agencies in order to reach space.

The world's first private spacecraft is an impressive feat of engineering marked by simplicity of design that, on closer inspection, doesn't look like it should fly into space. The interior is sparse and devoid of the myriad switches, dials, and toggles that crowd the Space Shuttle flight deck. There are a few low-tech levers, pedals, and

Table 1.1. The specs on SpaceShipOne.

SpaceShipOne	
Crew:	1 pilot
Capacity:	2 passengers
Length:	5 m
Wingspan:	5 m
Wing area:	15 m^2
Empty weight:	1,200 kg
Loaded weight:	3,600 kg
Powerplant:	1 × N20/HTPB SpaceDev Hybrid rocket motor, 7,500 kgf I$_{sp}$
Burn time:	87 seconds
Aspect ratio	1.6
Performance	
Maximum speed:	Mach 3.09 (3,518 km/h)
Range:	65 km
Service ceiling:	112,000 m
Rate of climb:	416.6 m/s
Wing loading:	240 kg/m^2

buttons that suggest the vehicle is designed to fly, but the austere design doesn't exactly scream "space". Clearly, SS1 (Table 1.1) is a very different spacecraft from all that have gone before, a mark of its uniqueness being its use of a hybrid rocket motor, which uses a mixture of solid fuel and laughing gas.

"WhiteKnight is taxiing" crackles over the public address system, an announcement that is followed shortly after by the sound of high-pitched jet engines that mark the arrival of the gleaming white carrier aircraft with SS1 slung tightly underneath. WhiteKnight and SS1 take off from Runway 30 at 06:47 local time, followed by two prop-driven chase planes, an Extra 300 and a Beechcraft Starship, which will follow SS1 during its one-hour ride to separation altitude, giving spectators plenty of time to grab another bagel and a coffee.

"Three minutes to separation" comes the announcement. Spectators scan the sky searching for the thin white line that is SS1. At the drop altitude of 14,000 meters, SS1 is released and dropped like a bomb above the Mojave airport. Falling wings level, pilot and soon-to-be private sector astronaut, ex-Navy Test Pilot, Brian Binnie, 51, trims SS1's control surfaces for a positive nose-up pitch and fires the rocket motor, boosting the spacecraft almost vertically. "It looks great," says Binnie as he rockets upwards at Mach 3. Within seconds, SS1 is gone, trailing a white line of dissipating white smoke. SS1 accelerates for 84 seconds, subjecting Binnie to three times the force of gravity as it races toward 45,000 meters altitude. The engines shut down and SS1 continues on its ballistic trajectory to an altitude of 114,421 meters. A loud cheer goes up from the spectators on the ground, who are following the proceedings on a giant screen, each of them euphoric with the realization that high above them, right there in the sky is a privately developed spacecraft that may one day carry them into space. High in the sky, with his spacecraft's rear wings feathered to increase drag on re-entry, Binnie prepares to bring SS1 back to Earth. On the ground the spectators wait, spellbound, straining to hear the double sonic boom that will announce SS1's return to the atmosphere. Seconds later the unmistakable sound announces that SS1 is on her way back from her historic mission, her signature shape descending to Earth in circles. Binnie guides SS1 gently back to Earth, gliding the spacecraft back to a perfect touchdown on the runway like any other aircraft. He has just become the 434th person to fly into space. Welcoming him enthusiastically on landing are 27,500 spectators, Microsoft's co-founder, Paul Allen, who helped finance the project, Burt Rutan, SS1's designer, and Peter Diamandis, chairman of the X-Prize Foundation. Private spaceflight has just become a reality. But this is just the beginning.

> "Today we have made history. Today we go to the stars. You have raised a tide that will bring billions of dollars into the industry and fund other teams to compete. We will begin a new era of spaceflight."
>
> Peter Diamandis, co-founder of the X-Prize Foundation, shortly after SpaceShipOne landed

> "It's a fantastic view, it's a fantastic feeling. There is a freedom there and a sense of wonder that—I tell you what—you all need to experience."
>
> Test pilot, Brian Binnie, describing his record-breaking trip

WHAT HAPPENED NEXT

SS1 was unveiled at the Smithsonian Institution's National Air and Space Museum on October 5th, 2005 in the Milestones of Flight Gallery and is now on display to the public in the main atrium between the Spirit of St. Louis and the Bell X-1. The whole project cost less than U.S.\$25 million, which is about the same amount that NASA spends every day before lunch, or less than a decent contract with the NBA! The price-tag is one of the most important aspects of SS1's flight since it finally demonstrated that passenger space flight travel, contrary to what was widely believed, really can be achieved at low cost. Also, although SS1's flight was suborbital, it gives a strong indication that the cost of orbital spaceflight can be similarly reduced.

> "I absolutely have to develop a space tourism system that is at least 100 times safer than anything that has flown man into space, and probably significantly more than that."
>
> Burt Rutan, shortly after SS1's landing

Shortly after the celebrations, Richard Branson, chairman of Virgin Atlantic Airways, announced that he will invest U.S.\$25 million in a new space venture to be called Virgin Galactic, a project that will license Rutan's Scaled Composites SS1 technology for commercial suborbital flights starting at U.S.\$200,000. The commercial flights will fly passengers in a spaceliner to be named SpaceShipTwo that will be operational in late 2009. For Branson, this venture will be different from any other that his Virgin group has been involved with. His travel business, cell phone company, and funky record business are all enterprises that have kept the champagne flowing, and kept Branson in the headlines, but until the flights of SS1, no Virgin business has ever had the potential to change the world. Virgin Galactic will be the world's first off-planet private airline no less, fielding a fleet of five spaceships by the end of the decade. The price tag for the whole venture is U.S.$121.5 million, or about half the price of a single Airbus A340-600, of which Virgin recently ordered 26.

"It may take decades. It may take 50 to 100 years. But it's going to lead to a new industry."

Dennis Tito, California millionaire and world's first paying space passenger

LESSONS LEARNED FROM SPACESHIPONE

The significance of the triumph of SS1 and its galvanizing effect upon the nascent space tourism industry illustrates some important lessons that many who have been accustomed to the government-bankrolled ventures such as the International Space Station may have forgotten.

First, SS1 represents a paradigm shift. For far too long the public associated space with government programs and assumed that travel into space was simply too expensive for the private sector. The reason for this common misperception was due to the government-sheltered monopoly that is NASA and its long tradition of suppressing vital private sector innovation. Now, thanks to the successful flights of SS1, Burt Rutan and Paul Allen have demonstrated not only that there can be a spaceflight revolution by initiating entrepreneurial competition, but that there can also be a free-market frontier. Other members of the embryonic commercial space crowd such as Jeff Bezos' super-secret Blue Origin, Rutan's neighbor XCOR, and European contenders EADS Astrium and Starchaser are each planning on being operational within the next five years. Commercial space tourism will definitely not be a monopoly!

Second, the flights of SS1, in winning the U.S.$10 million X-Prize, demonstrate clearly the motivational power of profit, although it has to be said that Scaled Composites was helped significantly by the deep pockets of Paul Allen. However, although the cost of the SS1 venture was U.S.$25 million and the prize was only U.S.$10 million, the real profit will be a long-term one when fare-paying passengers start to fly into space. Historically, cash prizes have done much to fuel the development of civil aviation. With Robert T. Bigelow's offer of a U.S.$50 million prize for the first team to develop an orbital equivalent of SS1, this tradition seems to be just as strong a motivator as it was when Charles Lindbergh won the U.S.$25,000 Orteig prize in 1927 for becoming the first pilot to fly nonstop across the Atlantic.

Third, SS1 demonstrates the power of pride. Scaled Composites and several other teams that were racing to win the prize struggled with limited resources and meager funding to develop new, innovative, and often ingenious ways of flying into space. By the manifestation of their creativity and despite great engineering and technical challenges, they took, and continue to take, great strides toward the reality of a space tourism business, as will be described in Chapter 2.

Finally, SS1 reminds everyone of the power of competition. The 20 teams who competed against each other for the X-Prize generated the dynamism of free enterprise that just simply doesn't happen in government-funded endeavors. The competition to fly into space demanded of the teams that they couldn't just offer an adequate product, especially when the product their competitors offered might be an excellent one.

Companies hot on the heels of SS1 include Canadian-based the Da Vinci Project, which had been considered SS1's closest competition. Da Vinci team leader Brian Feeney anticipates his company's XF1 craft, modeled on NASA's X-36 prototype, could be flight-tested in 2008. California-based XCOR plans to field a suborbital spacecraft called the Xerus. Rocketplane Kistler and its Rocketplane XL rocketized Learjet is scheduled to begin service in 2009, the same year that Virgin Galactic plans to start commercial operations.

Orbital momentum is also building in the fledgling private space industry, as billionaires have channeled independent fortunes into the cause of opening up space. In June 2006, Las Vegas hotel mogul, Robert Bigelow launched the first of his orbiting habitats, Genesis I, which was followed a year later by, you guessed it, Genesis II. Bigelow, a modern Howard Hughes is an adherent of the frontier formula:

$$\text{transportation} + \text{destination} = \text{habitation} + \text{exploitation} + \text{industrialization}.$$

As an owner of a private space company, Bigelow intends to make money on his investment by leasing his habitats to industry and to countries without a manned space program who want to fly their own astronauts. Beyond that, he has plans to go to the Moon and Mars.

Elon Musk's SpaceX has already flown two test articles of its Falcon 1 rocket, a precursor to its Falcon 9, which may in the near future ferry passengers to Bigelow's orbiting habitats. Jeff Bezos, the Amazon.com and Blue Origin founder is pursuing his own private space dream at a commercial spaceport in West Texas. Bigelow, Bezos, Musk, and company have their own money, their own business models, and the ability to finance what they are doing without any government help. In going it alone, the new "entreprenauts" have forced the governments and space agencies to face difficult questions, one of which is how to justify spending billions of dollars of taxpayer's funds on a space transportation system that is less efficient than one that the private space industry can develop for its own purposes? The question may be a difficult one for government space leaders to answer because historically the White House and Congress have been driven by the power of traditional aerospace lobbying and the need to maintain political constituencies, a situation that is still reflected by new White House space transportation policy. In the long run, the aerospace giants and possible soon-to-be dinosaurs of the aerospace industry, such as Boeing, Lockheed Martin, and Northrup Grumman will not be doing their stockholders any favors by clinging to a dying market, when all around them a frontier-based industry is expanding in leaps and bounds.

THE FUTURE

The concept of space tourism is not a new one. Following the Moon landings and the birth of the Space Shuttle program many people assumed it was merely a matter of time before they would be able to buy a ticket into space. The problem was the cost, with an average Space Shuttle mission priced at more than U.S.$400 million, which, if you divide that between a nominal crew of seven astronauts, equates to a ticket price of almost U.S.$60 million! With the loss of the Space Shuttle Challenger in 1986, the dream of space tourism was forgotten by many, until space entrepreneur, Peter Diamandis revived it with the launch of the X-Prize, a race that attracted aircraft designers from around the world. Burt Rutan was the first to sign up. Since the X-Prize was launched, a few people have already flown into space as paying passengers on the Russian spacecraft Soyuz, but for them the fare was as steep as the ascent to orbit. For the privilege of becoming a spaceflight participant, Dennis Tito, Mark Shuttleworth, Anousheh Ansari, Charles Simonyi, and Greg Olsen each paid U.S.$20 million, a fare that is clearly out of the reach of most people. However, thanks to SS1 the pulse of public interest in flying high above the Earth is being felt more and more as space tourism companies such as Rocketplane and SpaceX set ticket prices for their own versions of SS1. SS1 clearly helped the personal spaceflight industry turn a corner, although initially it will be a niche market that caters to people with a strong interest in space coupled with a desire to be among the first commercial space tourists and a wallet deep enough to pay a ticket price that is north of U.S.$100,000. This niche market will sustain the industry until the next generation of space vehicles are developed that will help bring the cost down to below U.S.$50,000, and eventually in the U.S.$10,000 range.

Beyond breaking the records and winning the X-Prize, the flights of SS1 gave life to the concept that is at the heart of the pro-space tourism movement; namely, that space is a place and not a program. By capturing the X-Prize, SS1 drove home the fact that space is open to all those who have the capabilities and drive to go there and, in demonstrating what a private space company with the right stuff can do, opened the door to a whole new industry that now has the chance to rise up and truly begin to open up the final frontier.

2

Suborbital company profiles, technology drivers, and mission architecture

What's the fastest way to become a commercial space millionaire?
Start as a commercial space billionaire.

Elon Musk, CEO, SpaceX

Those with even a rudimentary understanding of space appreciate the colossal economic barriers to entering the commercial launch business of space. The list of companies that have tried and failed to achieve suborbital flight is a long one, but this has not prevented those with lofty ambitions from trying to break into the commercial suborbital launch club.

It is important that anyone planning to spend a significant amount of money on a vacation as complex as spaceflight be informed of the advantages and disadvantages of the experience. Furthermore, for an activity with as many inherent risks as flying into space, it is equally important to be able to evaluate and understand the dangers involved. To that end, this chapter provides you with an introduction to the profiles, technology, and mission architectures of the suborbital companies likely to be offering tickets into space within the next five to ten years. Also described is an assessment of the risks of suborbital flight and which questions you should be asking pertaining to training, qualifications, medical standards, and choice of location. By the end of this chapter you should have sufficient information to make an informed choice regarding your operator.

SUBORBITAL FLIGHT RISKS

"No matter how much effort we put into safety, there is a chance of an accident."

Alex Tai, Chief Operating Officer, Virgin Galactic, speaking at the *International Symposium for Personal Spaceflight in Las Cruces, New Mexico, U.S., October 19, 2006*

The risks of a nominal suborbital flight for a healthy person are minimal, but that does not mean they do not exist. As aviation history has taught us all too often, sooner or later an anomalous situation will occur and despite what you may think, the policies and procedures for abnormal spaceflights and emergency procedures are not regulated to the same degree as for commercial aviation. To allow you to assess the risks of what will most likely be the most expensive holiday you will ever take, this section explains how the space tourism industry is regulated.

The role of the Federal Aviation Administration

The United States Federal Aviation Administration's Office of Commercial Space Transportation (usually referred to as FAA/AST) is the agency that approves any commercial rocket launch. The FAA/AST [1] requires that rocket manufacturers and launches comply with specific regulations to indemnify and protect the safety of people and property that may be affected by a launch. AST is a regulatory agency responsible for the licensing of private space vehicles and spaceports within the United States. The role of AST, therefore, is very different from the role of NASA, which is classified as a research and development agency of the United States Federal Government, a status that requires it to neither operate nor regulate the activities of the commercial space transportation industry. However, NASA can, and often does, launch satellites developed by private companies. According to AST's legal mandate under regulation 49 U.S.C., Subtitle IX, Chapter 701, *Commercial Space Launch Activities* [2], the agency has a responsibility to

- Regulate the commercial space transportation industry, only to the extent necessary to ensure compliance with international obligations of the United States and to protect the public health and safety, property, and national security and foreign policy interest of the United States.
- Encourage, facilitate, and promote commercial space launches by the private sector.
- Recommend appropriate changes in Federal statutes, treaties, regulations, policies, plans, and procedures.

One of AST's responsibilities is the licensing of rockets, the regulations for which can be found in 14 CFR, Chapter III [3]. The space tourism launch operator therefore has to comply with one of two sets of regulations, depending on whether the vehicle is an Expendable Launch Vehicle (ELV), in which case compliance must be in accordance with Part 415, or a Reusable Launch Vehicle (RLV), in which case compliance must be in accordance with Part 413. The launch site used by your operator must also comply with specifications laid down in AST's Part 420 regulations [4].

AST's approach to ensuring safety when licensing launch operations is to require that operators implement a three-phase approach that includes *quantitative analysis*, *a system safety process*, and an *assessment of the restrictions of operating*.

Quantitative analysis

One aspect of the quantitative analysis operators must perform is a calculation of the probability of casualty to any and all groups of people within the maximum dispersion of the vehicle. This calculation is called the Expected Casualty Calculation (ECC), which is an estimate of the risks that can be attributed to the vehicle. The estimate describes the containment of the rocket, which means that there are no people or property located within the maximum range of the vehicle. However, due to the various failure modes of the vehicle, the various locations of people, and the various means by which people can be injured, this estimate is very complex. Before a launch the operator must also be able to provide to AST an estimate of containment, which involves an assessment for each estimate of risk and what the GO/NO GO criteria are for each. The GO/NO GO criteria will include an assessment of the effects of direct impact by the vehicle, the effects of blast overpressure, toxic clouds emitted as a result of a launch-pad explosion, and various other malfunctions. If an operator submits an "unknown" against any of the criteria, then the AST response will be simply to refuse that operator permission to launch.

Safety system process

The safety system process is a little more complex to calculate due to the fact that new vehicles simply do not have the history to demonstrate reliability, which means that any assessment uncertainty is highly substantial. To assist in the calculation, various hazard analysis methods are used by an operator to complete the safety system process. These means of determination include methods such as the Failure Modes and Effects Analysis (FMEA) and top-down and bottom-up analyses. Each method investigates failure modes of the vehicle, its systems, external hazards, and any hazard that may threaten public safety. Based on an assessment of these hazards, mitigation measures are first developed, then verified by AST and finally implemented in the form of GO/NO GO criteria on launch day.

In addition to AST requiring operators to adhere to their particular set of GO/NO GO restrictions, operators must also be compliant with operating restrictions described in the Code of Federal Regulations [5]. A good example of such a restriction is one that prevents collisions with other manned spacecraft such as the Space Shuttle.

Despite the regulations and the requirement to comply with a myriad of threat assessments and risk analyses, one of the principles that operators rely on is the principle of informed consent. In other words, operators recognize that spaceflight is risky and they understand their passengers know that, but, from the operator's perspective, as long as passengers assume the risks by signing an informed consent form, the problem, at least from the operator's perspective, is largely taken care of. Unfortunately, if and when an accident does occur in the nascent space tourism industry such an attitude may be revisited, and revisited intensively!

Now that you understand the basic process of risk analysis, threat assessment, and how they are regulated, it is time to think about what questions you should be asking to determine your choice of operator.

> "I don't need that publicity anymore and I don't need that risk."
>
> Buzz Aldrin, the second human to set foot on the Moon,
> responding to an offer to fly on a spacecraft designed for tourists

Vehicle type

Given the complex technologies involved, the numerous failure modes of dozens of systems, and the probabilities of occurrence, it is difficult to define survival likelihood for a vehicle. For example, in selecting an operator you should consider the technologies and the mission architectures as these will have a bearing on how resistant the vehicle is to failure and catastrophic loss. One question you should ask is whether your operator's RLV is launched vertically (vertical takeoff, or VTO) or horizontally (horizontal takeoff, or HTO). In the case of a VTO RLV, for example, if a motor fails or is shut down during the first few seconds of flight, then the vehicle will be lost.

An HTO RLV, on the other hand, may suffer a motor shutdown during the first few seconds of flight and may still survive by conducting a runway abort or a go-around procedure.

Engine configuration

Another question you should ask is whether the vehicle uses an engine cluster or instead relies on just one engine. In a VTO RLV using a cluster of engines an engine-out capability (in which one engine fails) may prevent catastrophic failure. The bottom line when it comes to a multi-motor cluster configuration is that the potential for a motor failure is greater than for a single motor of similar reliability. Another way of explaining this is to use simple statistics. If your operator has tested a rocket motor and determined it to be 99.9% reliable for a suborbital flight profile, then the probability of a motor failure during a single suborbital flight is 0.1% (1/1,000). Conversely, if another operator uses a configuration of five motors with the same degree of reliability, then the probability of a motor failure during a single suborbital flight is 0.5% (1/200). Therefore, if you decide to choose an operator whose vehicle incorporates a multi-motor cluster configuration, you should perhaps ask which motor shutdown scenarios and contingencies feature in the design in the event of an off-nominal situation.

Human factor elements

Once you have determined the engineering risks you should now evaluate the human factor elements in the vehicle design. In other words, during the various phases of flight, what escape possibilities exist for each phase? An examination of the current space tourism vehicles reveals that few have any emergency escape capability or launch escape systems. The main reason for this is the redundancy of such a feature during almost the entire suborbital flight regime. For example, if a vehicle was to suffer a very low altitude abort the fully fueled vehicle together with its crew and passengers would be lost in a fireball which would render even the most powerful escape system ineffective. In such a scenario a vehicle would require an expensive escape tower such as was fitted to the Apollo system. However, for the initial launch and early ascent phase of flight, a launch escape system, such as the one featured in Starchaser's vehicle, may prove an attraction to some passengers who like a little extra insurance!

Once you have survived the first phase of low-altitude flight, you enter the intermediate phase during which you will be accelerated to speeds of between Mach 2.5 and Mach 3, a flight phase that clearly renders any current ejection system as effective as a deflated airbag! If you decide to fly with an operator that uses an HTO RLV, then there may be some merit in the use of an ejection system, especially during the climb-out and intermediate (subsonic) phases of flight. However, once a vehicle climbs beyond the 0.9 Mach envelope, ejection systems are of little value, although there has been one incident in which a pilot survived a bailout at speeds in excess of Mach 1. On January 25th, 1966, Lockheed test pilot, Bill Weaver, was flying the SR-71 Blackbird (SR-71A 64-17952) from Edwards Air Force Base in a series of maneuvers designed to reduce trim drag and improve high-Mach cruise performance. While conducting a programmed 350 bank turn, Weavers plane experienced an unstart on the right engine, a phenomenon that results in a shock wave being expelled forward of the engine inlet, and which caused the aircraft to roll to the right and pitch up, before gradually disintegrating, knocking Weaver unconscious. Although Weaver didn't initiate an ejection, he separated from the aircraft and regained consciousness to find himself tumbling high above the ranch lands of northeastern New Mexico.

"I blacked out. I thought I was having bad dream. This couldn't possibly be happening. If it had really happened, I couldn't survive."

Bill Weaver, SR-71 "Blackbird" pilot describing his thoughts
as his aircraft broke up at Mach 3.18 (3,200 km/h) while flying at 23,600 meters

Other potential occurrences include the possibility of a rapid or explosive decompression as a result of a motor explosion, which may cause engine fragments to penetrate the spacecraft. The risk of cabin depressurization at altitude will determine if your operator equips you with partial or full pressure suits. For example, based on the images of suits that will be worn by Virgin Galactic passengers, the risk of cabin depressurization is calculated as almost non-existent since Virgin's flight suits do not have the capability to deal with a loss of cabin pressure.

Other emergency scenarios you should investigate include the operator's ability to deal with contingencies such as post-abort water landings, and the training provided to passengers to deal with this and other emergencies, such as fire and emergency egress procedures.

Another important piece of information that will affect safety is the type of rocket engine that will propel the vehicle and its passengers. Let us take a look at the different types of rockets and the risks associated with each.

Rocket motors

The three types of rockets that will propel your vehicle include liquid propellant, solid, or hybrid. Of these, solid rocket motors are probably the least safe since they do not have an "off" switch nor are they capable of being throttled, although this risk factor didn't prevent NASA from choosing such a system for the Space Shuttle, and look what happened there! Once these rocket motors are ignited they simply burn and continue to burn until all rocket fuel is exhausted. Second, the nature of ammonium perchlorate, the toxic chemical usually used in solid-fueled rockets is extremely hazardous and represents an environmental contamination problem to its users. Another problem is the change in the center of gravity as fuel is burned which, in the case of a solid-fueled rocket, would make most high-performance vehicles unflyable, since the propellant mass represents such a large fraction of the load. For example, Xerus, the suborbital vehicle that XCOR intends to fly, is winged and therefore must be able to fly as a glider whether it is full, empty, or during an abort. Finally, because solid rocket motors create such severe shock and vibration environments they are not very space tourist–friendly.

Hybrid rocket motors have several advantages compared with solid rocket motors, one of which is the claim they are inherently safer, although you should be aware that both NASA and MSFC have experienced case ruptures on tests of large motors that would have destroyed a manned vehicle. After considering solid rocket motors and hybrid rocket motors you are left with liquid propellant motors. XCOR, which is a company that specializes in the design and manufacture of rocket engines, has conducted more than 2,000 engine runs of liquid propellant rocket motors without a single incident of a motor exploding or bursting.

Choice of spaceport

A spaceport is a site for launching spacecraft and usually includes several launch complexes and runways for takeoff and landing. Spaceports used by NASA, for example, include Cape Canaveral located in Florida, and Vandenburg located in California. These government launch complexes feature what would be considered rather austere facilities for future spaceflight participants, which is why the space tourism industry is constructing its own spaceports. For example, Mojave Spaceport, also known as the Mojave Airport and Civilian Aerospace Test Center, is a facility licensed in the United States for horizontal launches of reusable spacecraft such as SpaceShipOne, which conducted the first privately funded suborbital flight. Because of its proximity to airspace restricted from ground level to unlimited height and a supersonic corridor, Mojave is also the preferred location of several other private space companies such as XCOR Aerospace, Orbital Sciences Corporation, and Interorbital Systems. Another major U.S. spaceport is the Clinton–Sherman Industrial Airpark, also known as the Oklahoma Spaceport, a newly authorized spaceport located in the western part of the state and home to Rocketplane Kistler, an umbrella enterprise comprised of two private aerospace firms: Rocketplane Limited, Inc., and Kistler Aerospace.

In deciding which spaceport to choose you may want to consider proximity to major transportation corridors, accommodation at the spaceport, and for families and friends, the location of tourist attractions.

COST OF TRAINING AND FLIGHT

The cost of a suborbital flight advertised by private space companies such as Virgin Galactic and Starchaser Industries is in the order of U.S.$200,000. When it comes to choosing an operator, price is obviously an important consideration and, just like any Earth-based vacation, there are several important questions you should be prepared to ask. For example, although it is a given that the advertised price will include preflight training and the flight, what extras are included? Does the price include the spacesuit that you will be wearing during your flight? If you decide to complete the training separately, how long will the training be valid for?

SUBORBITAL COMPANIES

Now that you are aware of the risks of suborbital flight it is time to evaluate the companies that are offering the experience of a lifetime. The following sections profile nine suborbital companies and provides you with details regarding the development of each company, its technology drivers, vehicle design, and mission architecture. There is no attempt to rate the safety of these companies nor is there any attempt to rank them in preference of location, vehicle design, or any other feature associated with the company. The intention is to provide you with as much information as possible to allow you to make an informed decision regarding your choice of operator based on facts that should be considered by a spaceflight participant.

VIRGIN GALACTIC

www.virgingalactic.com

Profile

In July 2005 Burt Rutan's Scaled Composites and Richard Branson's Virgin group signed an agreement to form The Spaceship Company, an aerospace company with the sole purpose of constructing a fleet of suborbital spaceships. The Spaceship Company has contracted Scaled Composites for the research, development, testing, and certification of SpaceShipTwo (SS2), which will be the first of five such spacecraft that will take four passengers at a time to the edge of space. The customer for the Spaceship Company is Virgin Galactic, which is the space tourism company founded following the successful flight of SS1. Virgin still faces considerable hurdles in achieving their goal of commencing flight operations by the end of 2009, not least of which are the administrative safety regulations of the FAA, overcoming the fire that killed three workers in July 2007, and dealing with what is still complex technology.

Virgin Galactic ticket sales for the high-priced U.S.$200,000 suborbital seats are tallying up, and for those who want to be among the first 100 space tourists the opportunity has long since passed. The "chosen few" include celebrities, scientists, and even CEOs of rival space tourism companies such as Elon Musk, founder and CEO of SpaceX. Among Virgin's first customers is actress, entrepreneur, and thrill-seeker, Victoria Principal, previously famous for her role a Pamela Ewing on the hit television series *Dallas*. She will be joined by movie director, Bryan Singer, responsible for movies such as *X-Men* and *The Usual Suspects*, who was inspired by the idea of traveling into space while shooting a shuttle accident for the movie *Superman Returns*. British astrophysicist, Stephen Hawking, who recently flew on a parabolic flight during which he experienced 20-second bursts of weightlessness will join fellow scientist James Lovelock, proponent of Gaia Theory, who will be 90 years old when he flies. The first hundred to fly also include non-celebrity space fans such as George and Loretta Whiteside who plan to celebrate their honeymoon during their flight, although 3 minutes is quite a short time for most people! George Whiteside is the executive director of the National Space Society while his wife is in charge of an organization called Space Generation. Another category of customer will include, inevitably, extreme sports fans such as real estate developer and co-owner of the Los Angeles Lakers, Edward Roski Jr., who has climbed Mount Everest, cycled across Mongolia, and visited the *Titanic* onboard a chartered submersible in 2000. William Shatner was offered a free ride on the inaugural space launch. but the actor who played Captain Kirk in the hit science fiction series *Star Trek*, turned it down, saying: "I don't mind going up, but I need guarantees I'll definitely come back."

Technology

The design of SS2, which will be named VSS Enterprise in honor of *Star Trek*'s fictional Starship Enterprise, was completed in late 2005, but, unlike several other space tourism companies that publish all manner of technical details on their websites, there is scant information regarding the details of Rutan's latest vehicle. However, the vehicle will be unveiled, prior to the start of flight-testing, in late 2007, although this date may be pushed back due to the accident in July 2007 that claimed the lives of three Scaled Composites employees. It is known that SS2 and its new carrier aircraft, Eve (WhiteKnightTwo), will be approximately three times the size of SS1 and the original WhiteKnight, and will approximate the same dimensions as a Gulfstream V business jet, which stands 1.8 m high and 2.2 m wide.

Mission architecture

The mission architecture for Virgin Galactic's flight can be best appreciated by viewing a computer-generated depiction of a hypothetical flight at the YouTube website titled "Let the Journey Begin". The spacecraft will be capable of carrying six passengers and two pilots. VSS Enterprise will take off from the Mojave Spaceport in California while attached to the mother ship, WK2, until the configuration reaches 50,000 feet altitude, at which the countdown to release VSS Enterprise from the WK2 will commence. Once the countdown reaches zero, VSS Enterprise will drop from the WK2 and its rocket motor will ignite. In a matter of seconds passengers will be traveling at more than three times the speed of sound as VSS Enterprise heads toward space. At engine cutoff, passengers, who by now will be newly minted spaceflight participants, will experience three minutes of microgravity and will be able to perform aerobatic maneuvers or simply stare out of the windows and admire the view.

Spaceport and tourist attractions

The first Virgin Galactic passengers will fly from Mojave Airport, site of SS1's historic flights, but after 2010 Virgin will move its base of operations to the Southwest Regional Spaceport complex, located in Upham, about 70 km northeast of Las Cruces, New Mexico. Located in a remote mountain landscape not far from the White Sands Missile Range, the spaceport will feature mostly underground facilities, to preserve the desert landscape as well as to save water and energy. The chances of a flight being delayed due to weather will be unlikely, since the area has 340 days of launch-friendly weather each year.

ROCKETPLANE

www.rocketplaneglobal.com/index.html

Profile

In 1996, engineer and Air Force test-pilot instructor, Mitchell Burnside Clapp formed Pioneer Rocketplane with the intention of winning the U.S.$10 million Ansari Prize for the first private company to send a spacecraft into suborbital space and repeat the feat within two weeks. Clapp's design concept was based on a two-seat F-111–sized aircraft powered by turbofan engines and a kerosene/oxygen-burning RD-100 rocket engine that used an expendable upper stage. Unfortunately, due to financial problems the spacecraft never got beyond the design phase and the Ansari Prize was won by Burt Rutan's SS1, which, ironically, turned out to be great news for Pioneer Rocketplane! Overnight, the concept of private companies building rockets to fly into space was credible, and with credibility came investment capital in the form of current Chief Executive Officer, George French, a space enthusiast, who recognized Rocketplane's design as a viable means of reaching space at a profit. French, a previous winner of the NASA AMES Research Astrobiology Team Group

Achievement Award and the National Space Society's Entrepreneur of the Year Award, invested U.S.$13 million in the company and gave Rocketplane the cash it needed to finally progress beyond the drawing board. The next step was to find a leading aerospace engineer and French wasted no time in hiring David Urie, a veteran of leading-edge engineering projects such as the X-33 Venture Star, the Skunk Works spacecraft that was being groomed by NASA in the 1990s to replace the Space Shuttle. On his arrival at Rocketplane, Urie recruited his own team of engineers, among them Bob Seto, Learjet's program manager, Gary Lantz, an aerodynamics engineer whose previous job was with aircraft manufacturer Cessna, and several experts in the field of thermodynamics, trajectory and performance analysis, and systems engineering.

Other key Rocketplane personnel include President, retired Colonel Randy Brinkley, who previously served as NASA Program Manager for the International Space Station (ISS), as well as Mission Director of the Hubble Space Telescope Repair Mission. Vice President of Flight Operations and Services is Rocketplane's resident astronaut, Commander (U.S.N. Retd.) John Herrington. Herrington will be the pilot of the spacecraft that flies the winner of a Microsoft Corp.–sponsored competition (details of which can be found on *http://vanishingpointgame.com*) on their trip to space. Before that happens though, the company has to build the XP Spaceplane.

Technology

Urie and Lantz have an easier task ahead of them than engineers working for rival space tourism companies thanks to choosing the existing fuselage of a Learjet 25 instead of having to design one from scratch. Since the Learjet 25 was designed to tolerate a G loading of more than 3 G, its design parameters are similar to the G loads that the Rocketplane (which must tolerate 4 G) has to sustain during its rocket-powered ascent and descent. This means that the risk and development time Urie and Lantz have saved can be more profitably used in beefing up Rocketplane (Table 2.1 and Figure 2.1, see also color section) to ensure it will survive the forces of liftoff and re-entry. To do this Urie and Lantz have increased the structural strength of the tail, designed a strong delta-wing assembly, and covered the spacecraft with a protective coating that will protect Rocketplane from the atmospheric heating on re-entry. To ensure the spacecraft can withstand higher temperatures, however, the aluminum leading edges in the original Learjet 25 have been replaced with titanium. All these necessary upgrades have come with a cost, however, since the weight of the makeover has added 2,045 kilograms to the Learjet's original weight, which translates into a requirement for a much longer runway. To make room for all the kerosene and liquid oxygen that is required to launch four people into space, Rocketplane has cut a section from one Learjet and added it to another to add 50 cm to the overall length. This fuel will power the 16,400-kilogram thrust rocket engine that will be fitted into the Learjet's tail.

Rocketplane is powered by a turbojet propulsion system that consists of modified CJ610 engines that have a high thrust-to-weight ratio compared with other turbojets.

Table 2.1. Rocketplane by the numbers.

Length:	13 meters
Diameter:	1.5 meters
GTOW:	8,863 kg
Dry weight:	3,863 kg
Crew environment:	Cabin pressurized to 10 psi
Payload capacity:	431 kg
No. of jet engines:	2
No. of rocket engines:	1
Rocket propellant:	Liquid oxygen/kerosene
Nominal rocket thrust:	16,363 kg

Figure 2.1. Rocketplane (see also color section). Image courtesy: Rocketplane Global, Inc. Source: *http://www. rocketplane.com*

The engines are used during takeoff and the climb to rocket ignition altitude, after which they are shut down. They are ignited again on re-entry, but this is only as a safety measure, since they are not required for the descent. Just like the Space Shuttle, Rocketplane has a Reaction Control System (RCS) that allows the spacecraft to perform exoatmospheric maneuvers to ensure passengers have a good view through the windows and also to maintain correct orientation during re-entry.

Mission architecture

Rocketplane will fly to a minimum altitude of 100 km carrying three fare-paying passengers and one pilot. It will be capable of a 24-hour turnaround time between flights and will take off and land from the same runway. The flight will commence with a conventional aircraft horizontal takeoff from Oklahoma Spaceport using jet engines. Rocketplane will then climb and accelerate to a subsonic cruise level of 12,500 m, the altitude at which the pilot will switch to a liquid oxygen/kerosene rocket engine that will burn for 70 seconds. The rocket-propelled portion of the flight will result in a 3 G pull-up to suborbital altitude followed by a ballistic coast to mission altitude and a return to base in unpowered flight. At it highest altitude Rocketplane will provide at least three minutes of microgravity and thanks to maneuvering of its RCS, passengers will have a clear view of space and Earth.

Spaceport

Rocketplane flights will take off from the company's home base at Oklahoma Spaceport, home to one of the world's longest runways (4,115 meters) as well as assorted ground facilities. Located near the town of Burns Flat, near the exit for Highway 44, the spaceport was formerly known as the Clinton–Sherman Air Force Base, but has now been renamed the Oklahoma Space Industry Development Authority, with the mission of attracting companies to build and launch spacecraft from Burns Flat. Since the next closest major city to Oklahoma City is Amarillo, 400 km away, there is little in the way of tourist attractions for the families and friends of Rocketplane passengers so it remains to be seen what facilities the spaceport will provide.

SPACEDEV

www.spacedev.com/

Profile

Based near San Diego in Poway, California, SpaceDev was founded in 1997 by Jim Benson with the vision of providing reliable, low-cost access to space. SpaceDev's Chief Executive Officer, Mark Sirangelo, previously served as a senior officer of Natexis Bleichroeder, Inc., an international investment-banking firm that focused on the security and defence sectors. On November 16th, 2005, SpaceDev announced details of its suborbital vehicle that will be capable of carrying four passengers, using internal hybrid rocket motors similar to those used by SS1. On September 28th, 2006, Jim Benson stepped down as chairman and chief technology officer to start the Benson Space Company (BSC), a space tourism venture that will provide funds to SpaceDev for funding of its spacecraft. Later that year, on December 18th, BSC awarded SpaceDev a $330,000 Phase I study contract to develop the DreamChaser's rocket motor systems and development of the spacecraft.

Technology

Perhaps SpaceDev's best-known technology is its rubber-burning hybrid rocket motors that it designed and built for Scaled Composites, which powered SS1 to its historic X-Prize winning flights. The DreamChaser^{TM} will be a four-passenger vehicle that was originally based on the X-34 for NASA's Commercial Orbital Transportation Services (COTS) program, but has since been revised and is now based on the ten-passenger HL-20 Personnel Launch System developed by NASA, Langley. Since the vehicle is derived from a NASA design, SpaceDev have had the advantage of having the information gleaned from more than 1,200 hours of wind tunnel testing and the seven years that NASA spent developing the original HL-20. In developing DreamChaser^{TM} (Figure 2.2), SpaceDev is being assisted by NASA's Ames Research Center (ARC), with whom it signed a Space Act agreement in 2004, to explore designs for affordable suborbital spacecraft. In its present design SpaceDev's vehicle will use internal hybrid rocket motors and will launch vertically using synthetic rubber as the fuel and nitrous oxide for the oxidizer to make the rubber burn. Unlike conventional rocket motors, SpaceDev's hybrid equivalent is exceptionally safe as they do not

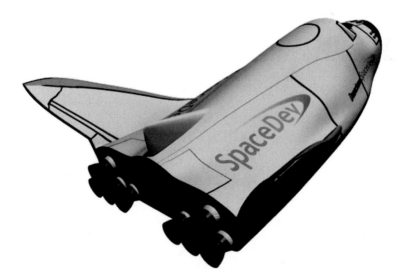

Figure 2.2. SpaceDev's DreamChaser^{TM}. Image courtesy: SpaceDev. Source: *www.spdv.info/wp-content/uploads/2007/02/dreamchaser1.jpg*

detonate, and for those who are environmentally conscious the motors do not produce toxic exhaust. Unlike several other motors being used by private space companies, SpaceDev is in the unique position that their motors have already been man-rated and have the advantage over conventional motors as they are both throttleable and restartable.

Mission architecture

In April 2007, SpaceDev announced that it had partnered with the United Launch Alliance to investigate the possibility of utilizing the Atlas V booster rocket as DreamChaser^{TM}'s launch vehicle. The spacecraft will be launched vertically from the ground and return to Earth by gliding back for a normal horizontal runway landing in a manner very similar to the Space Shuttle. In fact, SpaceDev personnel have flown the DreamChaser^{TM} in the Vertical Motion Simulator at NASA's ARC, and found that the flight characteristics were very similar to those of the Shuttle. The company plans to conduct multiple manned suborbital test flights by 2008, although this timeline is subject to the availability of adequate funding.

Spaceport and tourist attractions

A launch site has yet to be confirmed, although potential sites include the Mojave Spaceport in California, and the Kennedy Space Center.

XCOR AEROSPACE

www.xcor.com/

"Mojave is to the emerging space industry what Silicon Valley was to the computer industry. There is a concentration of like-minded businesses here like no where else in the world."

Jeff Greason, President and Co-founder, XCOR, and
Mojave Flight Test Center Tenant

Profile

XCOR Aerospace is a small, privately held California corporation located at the Mojave Spaceport and Civilian Test Center in Mojave, California. While the company may not have the deep pockets of some of its competitors it certainly has one of the most impressive teams. President and co-founder of XCOR Aerospace, Jeff Greason, holds 18 patents and was cited by *Time* magazine in 2001 as one of the "Inventors of the Year" for his team's work on the EZ-Rocket. Dan DeLong, XCOR's Chief Engineer, has extensive experience in the space industry, having served as lead engineer for Boeing's Life Support Systems internal research development program, as Principal Engineer at Boeing Defense & Space, and as an engineering analyst in the Life Support group that developed the ISS air and water recycling systems. XCOR's test pilots, Rick Searfoss and Dick Rutan, are perhaps the most experienced of any suborbital space company. Colonel Rick Searfoss's credentials include serving as Space Shuttle pilot on STS-58 and STS-76, and Space Shuttle Commander for STS-90, while Dick Rutan is famous for flying the Voyager aircraft around the world non-stop with Jeana Yeager in an aircraft designed by his brother, Burt Rutan, of Scaled Composites fame.

Much of XCOR's money has come from customer contracts and angel investors such as the Boston Harbor Angels, Esther Dyson, Stephen Fleming, and Lee Valentine. In April 2007 XCOR was also awarded a Small Business Innovative Research (SBIR) Phase I contract from the Air Force to design and analyze a rocket-powered vehicle capable of reaching 60,000 meters as part of the Air Force Research Laboratory (AFRL).

Technology

XCOR is currently engaged in developing its piloted rocket operations demonstrator aircraft, the EZ-Rocket, with plans to proceed toward a phased development of its next generation vehicle, the suborbital RLV, Xerus (Figure 2.3, see also color section). The EZ-Rocket is a manned technology demonstrator, which serves as a test bed for XCOR's future vehicles. It has been flown 26 times under rocket power and has enabled XCOR to more fully understand routine rocket-powered flight and operations. The EZ-Rocket has the ability to re-start engines in mid-flight and

Figure 2.3. The Xerus spacecraft is a single-stage suborbital vehicle capable of servicing three markets: microgravity research, space tourism, and microsatellite payloads (see also color section). Source: *www.xcor. com/products/index.html*

perform touch and goes, a feature that no other rocket-powered aircraft has. Interestingly, SS1 astronaut, Mike Melville experienced his first rocket vehicle when flying the EZ-Rocket.

XCOR now has a firm base of experience concerning rocket-powered flight and is currently involved in the design of the Xerus craft, a single-stage suborbital spacecraft that will be capable of providing service to markets such as space tourism and microgravity payloads. The Xerus is being designed to take off and land from a conventional runway without the need for a carrier aircraft such as the SS1 configuration. XCOR plans a flight test program of approximately 20 flights, each flight incrementally extending the operational envelope.

The Xerus program is still in the design phase but thanks to the company's experience with rockets and its decision to use a liquid propellant rather than solid or hybrid rocket engines, the vehicle promises a high degree of safety. Additional safety is assured by the fact that the motors are restartable and have had a total of almost 700 runs lasting nearly 3 hours. The motor that will power Xerus to suborbital altitude will be XCOR's third-generation 1,800 lb thrust liquid oxygen (LOX)/ kerosene engine.

Mission architecture

Xerus will take off from a conventional runway and burn its engines to propel the vehicle at a speed of up to Mach 4 to an altitude of 65 km, at which point inertia will coast the vehicle to its intended orbit of 100 km. During the suborbital phase the pilot will use the RCS to maneuver the vehicle and adjust attitude if necessary. Xerus will retain a propellant reserve that will enable the pilot to restart the engines or perform a go-around if required. Since the vehicle is winged it will return to Earth in the same way that the Shuttle does. Tickets for flights on Xerus are being sold through Space Adventures, priced at U.S.$98,000.

Spaceport

XCOR passengers will begin their trip into space at the Mojave Spaceport, also known as the Mojave Airport and Civilian Aerospace Test Center, located in Mojave, California.

EADS ASTRIUM

www.space.eads.net/company/eads-astrium

"We believe it is the will of human beings to visit space and we have to give them the possibility to do that."

François Auque, CEO of Astrium, in an interview with *BBC News*

BLUE ORIGIN

http://public.blueorigin.com/index.html

Profile

Founded in September 2000, Blue Origin is owned by Amazon.com founder, Jeff Bezos, and headquartered in a warehouse on East Marginal Way, a quiet street in Kent, a suburb of Seattle, Washington. Kent is the location of the company's research and development, whereas its testing and operations are conducted near the Guadalupe Mountains in Texas. Bezos, who has a net worth of $1.7 billion, has a 20-year plan that envisions his company developing near-Earth space not only for passenger travel but also for industry and commercial applications. To achieve his dream he has recruited an experienced team that includes Jet Propulsion Lab and Mariner probe veteran, Jim French, physicist Maclen Mervit, and aerospace engineer Tomas Svitek.

Technology

The New Shepard is a vertical takeoff and vertical landing (VTOVL) RLV that will comprise a propulsion module and a crew capsule capable of carrying three space-flight participants to space. It will take off vertically, fly a suborbital trajectory, and land in a vertical powered landing. The propulsion module will use high-test peroxide and rocket propellant–grade kerosene that will provide 104,545 kilograms of force at liftoff. As with several other private spacecraft designs, New Shepard will feature a low-thrust RCS that will permit exoatmospheric maneuvers. The short and squat crew capsule, which looks something like a scaled-up version of the Apollo Command Module, will carry its own solid rocket motors that will be used in an abort situation. Unmanned tests of Blue Origin's subscale demonstration New Shepard vehicle (Figure 2.7) began in November 2006, using a prototype vehicle, known as Goddard. Incremental flight-testing to the vehicle's planned 100 km altitude is to be conducted between 2007 and 2009 with a full-scale vehicle expected to fly in 2010.

Mission architecture

New Shepard will carry three crewmembers and will take off and land vertically under its own propulsion, using technology similar to that employed by McDonnell Douglas in its development of the DC-X, a vertically launched rocket developed for a missile-defense system. A typical mission will feature an almost vertical sub-orbital trajectory that will last two minutes, followed by a coast to an altitude in excess of 99,060 m (325,000 feet) before descending and restarting the vehicle's engines several thousand meters above the ground for a powered landing. The total time from liftoff to landing will be fewer than ten minutes. In the event of an emergency situation during liftoff the crew capsule will separate from the propulsion module using the solid rocket motors attached to the crew capsule. The crew capsule

Figure 2.7. Blue Origin's New Shepard vertical takeoff vertical landing vehicle. Image courtesy: Blue Origin. Source: *http://public.blueorigin.com/index.html*

will then be jettisoned and the propulsion module will attempt to land back at the landing site.

Spaceport

Blue Origin is constructing a privately owned space launch site in Culberston County, 25 miles north of Van Horn, Texas, which will include a vehicle-processing facility, launch complex, vehicle landing and recovery area, and spaceflight participant–training facilities.

ARMADILLO AEROSPACE

Profile

http://armadilloaerospace.com/n.x/Armadillo/Home

Based in Mesquite, Texas, Armadillo Aerospace is financed and led by rocket entre-preneur and pioneering computer programmer John Carmack who is perhaps best known for the immensely popular and highly addictive computer games "Quake" and "Doom". In the six years of the company's existence Carmack has spent more than U.S.$3 million investing in his dream of providing manned suborbital access.

Technology

Armadillo aim to provide the space tourism industry with a new generation of reusable launch vehicles based on mass-produced, modularized propulsion systems. The end product will be a computer-controlled, VTOVL system that will carry passengers to suborbital altitudes using liquid oxygen and ethanol-fueled engines. Since 2001 Carmack's company has conducted more than 100 rocket-powered test flights using different propellant combinations and engines. The launch vehicles that the company has tested have used an array of attitude control systems and several generations of control systems.

Mission architecture

Armadillo's piloted vehicle will comprise a cabin with a cluster of six to nine modules in a single stage. Details of the mission profile have yet to be defined but you can expect a similar flight regime to Blue Origin's VTOVL vehicle.

Spaceport

Armadillo Aerospace will launch from Oklahoma Spaceport.

DA VINCI PROJECT

Profile

www.davinciproject.com/

Based in Toronto, Canada, the da Vinci Project (DVP) was founded in 1996 by Brian Feeney whose background is in the disciplines of closed-loop life support systems and the development of life support solutions for aerospace and military applications. Feeney was originally inspired by the challenge of flying into space when reading an article about the X-Prize in 1996, and since then has dedicated his company to achieving that goal. The DVP team comprises a small army of 600 volunteers who dedicate their time toward the goal of the operational readiness of the company's suborbital vehicle, Wildfire MKVI.

Technology

Da Vinci's spacecraft, WildFire MKVI (Figure 2.8, see also color section), which has a diameter of 2 m (78 inches) and a length of 8 m (26 ft), is designed to carry three passengers in a one atmosphere–pressurized atmosphere. The launch system comprises a rocket and an unmanned reusable helium balloon, which will lift the rocket to ignition altitude. The single, pressure-fed hybrid rocket engine will be fueled with solid paraffin and liquid nitrous oxide providing a total thrust of 80,000 N (18,000 lbf). To enable the vehicle to perform atmospheric and exoatmospheric maneuvers it is fitted with an RCS that will use cold gas nitrogen.

Mission architecture

WildFire is designed to be air-launched from the world's largest fully reusable helium balloon at an altitude of 21,340 m (70,000 ft). The ascent duration to ignition altitude is projected to take between 90 to 120 minutes during which time the rocket will be tethered 250 m (820 ft) below the base of the balloon. On attaining ignition altitude a series of launch procedures will be initiated and a 120-second countdown sequence will commence. Shortly after ignition, passengers will be subjected to 3.5 G, which will represent the peak G loading during their 90-second trip to the main engine cutoff (MECO) altitude of 63,000 m. At this altitude the vehicle will be traveling at Mach 3.5 or approximately 1.19 kilometers per second. At 85,000 m the capsule will separate from the propulsion section, which will freefall to an altitude of 12,190 m at which point drogue chutes will deploy followed by a main chute deployment at 3,050 m. The capsule will continue to apogee, which is projected to be 115,000 m, the altitude at which passengers can expect to spend up to 3.5 minutes of microgravity. After re-entry, the capsule will follow the same recovery sequence as the propulsion section, but unlike the propulsion section the capsule will be cushioned by airbags.

Figure 2.8. The da Vinci Project's Wildfire MKVI vehicle (see also color section). Image courtesy: da Vinci Project. Source: *http://www.davinciproject.com/documents/Canadian_da_Vinci_Project_Team_Summary_Aug_2004_V6.).pdf*

Spaceport

The DaVinci Project has selected the town of Kindersley, as its first spaceport. The small town is located in Canada's Saskatchewan province and its selection was based on the launch-friendly weather, proximity of hotels, and flat stretches of land that are all elements conducive to the requirements of a spaceport.

The small steps taken by the private space companies described in this chapter will quickly evolve into giant leaps for affordable, commercial human spaceflight. Companies such as Virgin Galactic and Rocketplane are already selling tickets whereas Blue Origin and Astrium are still several years from being operational. For those who have U.S.$200,000 and are ready to fly into space the following chapter will describe the steps you must take to fulfill the medical and training requirements.

REFERENCES

[1] *Study of the Liability Risk-Sharing Regime in the United States for Commercial Space Transportation.* Aerospace Report No. ATR-2006 (5266)-1.
[2] *Commercial Space Launch Amendments Act, 2004.* Public Law 108-492. 108th Cong.
[3] Da Vinci X-Prize space project: mission analysis. *Proceedings of Third International Symposium: Atmospheric Re-entry Vehicles and Systems, Archachon, France, March 24–27, 2003.*
[4] *http://www.faa.gov/about/office_org/headquarters_office/ast/reports_studies/*
[5] *http://www.faa.gov/about/office_org/headquarters_offices/ast/licences_permits/htm*

3

Medical and training requirements for suborbital flight

Now that you have chosen an operator and paid your 10% U.S.$20,000 deposit, it is time to consider the medical and training aspects of your journey into space. To help you prepare, this chapter outlines the next steps a spaceflight participant must take.

The length of training will vary between operators as it will be determined by various factors pertaining to the vehicle and its systems. For example, your training will be largely defined by the technologies utilized in the design of the vehicle and whether the spacecraft has a launch escape system (LES), an emergency egress system, single-engine vs. engine cluster design, and engine-out capability. Some operators such as Virgin Galactic, which does not demand that its passengers wear a pressure suit and whose mission architecture does not offer an emergency egress, require you to train for only three days prior to your suborbital flight. Starchaser, however, due to its vehicle type with its LES and the necessity to train in the use of a pressure suit, advertises a two-week training schedule. Clearly, each operator will design its training according to the specific needs of its vehicle and mission architecture, which is why the syllabus described in this chapter is a generic suborbital training schedule, featuring most of the subjects you can expect to be delivered by a private space company.

TRAINING AND MEDICAL REQUIREMENTS FOR SUBORBITAL FLIGHT

The medical and training standards required of suborbital and orbital crew and passengers are stated in a Federal Aviation Administration (FAA) document titled *Human Space Flight Requirements for Crew and Space Flight Participants: Final Rule* [4], which your operator is obliged to follow. These requirements were established by the Commercial Space Launch Amendment Act (CSLAA) of 2004 [1] and include rules on crew qualifications, as well as training guidelines for spaceflight participants.

After you have registered for a spaceflight, your operator will send you a list of the FAA-certified flight surgeons in your area together with a file containing the medical standards required for certification of a spaceflight participant. Included in this file will be a medical history form, which you will be required to complete prior to your appointment with the flight surgeon. The medical standards that you will be expected to meet are outlined in Table 3.1.

Table 3.1. Medical standards for suborbital spaceflight participants [3].

Medical standard	Spaceflight participant
Distant vision	20/40 or better separately, with or without correction
Near vision	20/40 or better in each eye separately (Snellen equivalent), with or without correction as measured at 16 inches
Intermediate vision	No requirement
Color vision	Ability to perceive those colors necessary for safe performance in space vehicle
Hearing	Demonstrate hearing of an average conversational voice in a quiet room using both ears at 6 feet with the back turned to the examiner, or pass one of the audiometric tests below
Audiology (must pass either test)	Audiometric speech discrimination test score of at least 70% in one ear. Pure tone audiometric test, unaided, with thresholds no worse than: | Hz	500	1,000	2,000	4,000
---	---	---	---	---
Better ear	35 dB	30 dB	30 dB	40 dB
Worse ear	35 dB	50 dB	50 dB	60 dB	
Ear, nose, throat (ENT)	No ear disease or condition manifested by, or that may be reasonably expected to be manifested by, vertigo or a disturbance of speech or equilibrium
Pulse	Used to determine cardiovascular system status and responsiveness
Electrocardiogram	Not required
Blood pressure	No specific value stated. Recommended possible treatment if BP is on average greater than 155/95 mmHg
Psychiatric	No diagnosis of psychosis, bipolar disorder, or severe personality disorder

Medical standard	Spaceflight participant
Substance dependence or substance abuse	A diagnosis or medical history of substance dependence or abuse will disqualify you unless there is established clinical evidence, satisfactory to the flight surgeon, of recovery, including sustained total abstinence from the substance(s) for 2 years preceding the examination. Substance includes alcohol and other drugs (i.e., PCP, sedatives and hypnotics, anxiolytics, marijuana, cocaine, opiates, amphetamines, hallucinogens, and other psychoactive drugs or chemicals)
Disqualifying conditions	An SFP will be disqualified if they have a history of (1) diabetes mellitus requiring hypoglycemic medication; (2) angina pectoris; (3) coronary heart disease that has been treated or, if untreated, that has been symptomatic or clinically significant; (4) myocardial infarction; (5) cardiac valve replacement; (6) permanent cardiac pacemaker; (7) heart transplant; (8) psychosis; (9) bipolar disorder; (10) personality disorder that is severe enough to have manifested itself by overt acts; (11) substance dependence; (12) substance abuse; (13) epilepsy; (14) disturbance of consciousness without satisfactory explanation of cause; or (15) transient loss of control of nervous system function without satisfactory explanation of cause
Dental examination	Panorex and dental X-rays within prior 2 years
Psychiatric and psychological evaluation	Psychiatric interview and psychological tests.
Other tests	Drug screen. Microbiologic, fungal, and viral tests. Pregnancy test. Screen for sexually transmitted disease. Abdominal ultrasound.

Adapted from *FAA Guide for Aviation Medical Examiners* [3].

INFORMATION FOR THE SPACEFLIGHT PARTICIPANT MEDICAL CERTIFICATE

To ensure your medical examination is successful you should follow the procedures outlined in this section.

Issuance, denial, and deferral

There are three possible outcomes of your spaceflight participant medical examination, the first of which is issuance of your Spaceflight Participant Medical Certificate (SFPMC), in which case you can proceed to the next step.

The second outcome is that you will be denied certification, in which case the flight surgeon will issue a formal denial letter indicating you are medically disqualified from spaceflight. If this happens, do not despair, as you have 30 days in which to appeal the decision. If, within this period you are able to present medical information that confirms your disqualifying medical condition is safe, you may be granted a waiver. To do this you will need to be examined by an aerospace surgeon who specializes in your disqualifying condition. Once you have obtained this information you must submit it for review to the flight surgeon who will make a medical recommendation to your operator. Based on the flight surgeon's recommendation you will be either approved or denied permission to fly.

The third outcome is that you will be deferred. For example, there are certain medical conditions that are subject to the waiver process provided you are able to meet certain requirements. If you receive a deferral, the flight surgeon will defer your file to an aerospace surgeon who specializes in the deferral condition. If, after having been examined by the specialist you meet the waiver condition, you will be issued with the SFPMC. If you do not meet the waiver condition you will be issued with a letter of denial and will be required to take the same steps as for a denial.

The waiver process is a fair system that has been developed to ensure a consistent and correct management of those whose medical certification has been deferred or denied. Often, waiver requests involve little more than a review and endorsement before final approval. In certain, more complicated cases it may be necessary for the specialist surgeon to request a second opinion which may require a potentially time-consuming case review.

A complete review of waiver criteria is beyond the scope of this chapter, but an overview of the factors considered in granting a waiver may be useful to those who anticipate the need to submit to the process. When considering a waiverable condition, the flight surgeon must first make provision for flight and crewmember safety. Second, to be considered waiverable, any disqualifying condition must comply with the following criteria:

a. The condition shall not pose any possible risk for subtle incapacitation that may not be detected by the individual but would affect alertness and/or ability to process information.
b. The condition shall be resolved and non-progressive.
c. The condition shall not require treatment, which has the potential to compromise training and/or flight.
d. The condition shall not pose any inherent risk for sudden incapacitation.
e. The condition shall not pose any potential for jeopardizing the successful completion of a mission.

To ensure you have the best possible chance of being issued your medical certificate you should follow the guidelines in this section.

Considerations when scheduling a medical examination

Once you have scheduled a flight date you must complete the medical within a three-month window preceding your flight. If you think you have a medical condition that may disqualify you or require a waiver you should seek an appropriate specialist at this time.

Completing the medical forms

When completing the medical forms it is important you report prescription and non-prescription medicines you are using and you must answer all questions on the Report of Medical History Form truthfully! Also, be sure to check each item on the Report of Medical History Form, as failure to do so may delay the examination and certification. You must also report all visits to physicians within the last 3 years and be prepared to submit additional information to the flight surgeon if required.

Administration

To speed up your examination, you should bring documentation relating to any treatment for any condition(s) you have received in the last 5 years, the details of any surgery performed, and appropriate medical records that pertain to any treatments, hospitalization, injury, and conditions. If you wear glasses or contact lenses then you should bring these along as well.

Before the examination

Ensure you get a good night's sleep and avoid strenuous physical activity the evening prior to or on the morning of the examination. You should also avoid smoking and minimize your caffeine intake as either of these may affect your electrocardiograph (ECG) test. Be aware that medications such as decongestants often contain ephedrine that may compromise your cardiovascular assessment. When deciding on what to eat for breakfast on the day of the medical, steer clear of high-sugar meals since an excess amount of sugar in your bloodstream may cause an abnormal result in the urinalysis. Instead, eat a light meal with complex carbohydrates and proteins.

Once you have been issued with your SFPMC, the flight surgeon will forward the details to your operator and you can begin to prepare for your training. To help in your preparation your operator will send you the training manuals and training schedule that you will be following during your six days at the training facility. As the training will be rather intensive it is advisable to spend an appropriate amount of time reviewing the material so you are well prepared on arrival.

The training will be divided into a series of theoretical and practical modules over five days. The theoretical modules will include lessons introducing you to such subjects as the history of astronaut training, space physiology, suborbital trajectory, spacecraft systems and subsystems, parabolic flight, and G-tolerance training. The practical modules will be instructional sessions that will introduce you to basic survival skills, high-altitude indoctrination, centrifuge training, vehicle simulators, and parabolic flight training.

GENERIC SUBORBITAL TRAINING PROGRAM

The following sections describe a generic 21-hour suborbital training program designed to convey pertinent operational information to those embarking on a suborbital flight and to those interested in the process of the training of a spaceflight participant. The information described in each module is a synopsized version of what will be delivered by the instructors and is intended to provide you with an overview of the subject material.

DAY 1

Time	Event/Training module	Details	Location
08:00–09:50	Met at airport by operator representative	Escorted to your hotel. Attend to personal administration.	Operator Training Facility
10:00–10:20	Coffee break		
10:20–11:10	Administration	Submit documentation required prior to commencement of training	Operator Training Facility
11:20–12:10	Medical check	Final preflight medical check	Operator Training Facility
12:15–13:05	Lunch break		
13:10–14:00	Equipment issue	Collect additional training manuals, laptop, flight suit, mission patches, and flight case	Operator Training Facility
14:10–15:00	Meet and greet	Introductions to your instructors, pilot, and fellow crewmembers	Operator Training Facility
15:10–16:00	Facility orientation	Tour of the spaceport, mission control, and training facilities	Operator Training Facility
16:10–17:00	Vehicle orientation	Your first chance to see the vehicle that will take you into space	Operator Training Facility
17:10–18:00	Dinner		
18:00–20:00	Personal administration and preparation for training		

DAY 2

Time	Training module	Details	Location
08:00–08:50	*AA1* History of astronaut selection	Theory. PPT. Required reading: Section 1 of *SFP Manual*	Operator Training Facility
09:00–09:50	*SA1* Spaceflight theory. Rocket engines.	Theory. PPT. Required reading: Section 2 of *SFP Manual*	Operator Training Facility
10:00–10:20	Coffee break		
10:20–11:10	*VA1* Vehicle indoctrination session	Theory. PPT. Required reading: *Operator Vehicle Manual*	Operator Training Facility
11:20–12:10	*HA1* High-altitude indoctrination, Part I	Theory. PPT. Required reading: Section 3 of *SFP Manual*	Operator Training Facility
12:15–13:05	Lunch break		
13:10–14:00	*PA1* Space physiology	Theory. PPT. Required reading: Section 3 of *SFP Manual*	Operator Training Facility
14:10–15:00	*TA1* Survival training	Theory. PPT. Required reading: Section 4 of *SFP Manual*	Operator Training Facility
15:05–15:25	Coffee break		
15:30–16:20	*HA2* High-altitude indoctrination, Part II	Theory. PPT. Required reading: Section 3 of *SFP Manual*	Operator Training Facility
16:30–17:20	*GA1* G-Tolerance indoctrination	Theory. PPT. Required reading: Section 5 of *SFP Manual*	Operator Training Facility
17:30–18:20	Dinner		
18:30–20:30	Fly to NASTAR facility, Southampton, Pennsylvania		
21:00–22:00	Check in to five-star hotel near NASTAR		

MORNING OF DAY 2

History and overview of astronaut selection: Academic Module SA1

As of June 13, 2007, 460 humans from 39 countries have flown in space according to the Fédération Aéronautique Internationale (FAI) [4] guideline that defines space-flight as any flight that occurs at an altitude higher than 100 km (62 miles). Of those 460 space-qualified humans, 456 have reached Earth orbit or beyond and 24 have traveled beyond low-Earth orbit (LEO). Of the 460 who have flown in space, 19 have been killed on 5 space missions and another 10 in ground-based accidents.

The process of selecting the first U.S. astronauts began in 1959, when NASA requested the U.S. military provide a list of personnel who met very specific qualifications that included jet fighter experience, engineering training, and a height restriction of 5 feet 11 inches due to the limited cabin space in the Mercury capsule. Other selection criteria required the astronaut candidate to be younger than 40 years, in excellent physical condition, have a minimum of 1,500 hours flying time, and be a qualified test pilot. The initial candidate pool from the Army, Navy, Air Force, and Marine Corps totaled 600, from which NASA invited 110 for further testing.

The physiological and psychological tests administered to the first group of astronauts included isolation tests in which the candidate astronaut had to spend 3 hours in a soundproof room, a heat test that subjected the candidate to temperatures of 130°F for 2 hours, and a self-inventory 566-item questionnaire. After an exhaustive and elaborate search and selection process, NASA selected its group of 7 astronauts in April 1959. Another 9 pilot astronauts were selected in September 1962, and 14 more in October 1963, by which time the selection emphasis had shifted away from flight experience and more toward advanced academic qualifications. A year later, applications were taken from 400 candidates based on academic qualifications alone, of which 6 were selected in June 1965.

A major change in the selection process occurred in 1977 when NASA began its recruitment of the Space Shuttle era of astronauts, the first of which were selected in January 1978. To have a chance of being selected as a shuttle astronaut it was necessary to have at least a doctoral degree, a pilot's licence, scuba-diving qualification, and extensive professional experience in a field related to the field of manned spaceflight. From more than 5,000 applicants, 200 were selected for interview and, based on the interviews, 35 were selected as astronaut candidates and assigned to the astronaut office at the Johnson Space Center (JSC) in Houston, Texas.

Since the Shuttle era, NASA has not made many changes to the selection process, which still requires applicants to submit to an exhaustive batch of tests, which include extensive laboratory examinations of blood work, urinalysis, and stool analysis, neurologic evaluations such as electroencephalograms at rest and during sleep, and extensive radiographic examinations of sinuses and chest. Although most astronauts do not look forward to any of the tests, there are some tests such as the uncomfortable proctosigmoidoscopy which, if you happen to mention it to an astronaut, is guaranteed to turn them pale! Fortunately, you do not have to submit to this test in order to qualify for a suborbital flight!

NASA's Astronaut Candidate Training is delivered at JSC, and is comprised of classes on the subjects of basic science, mathematics, meteorology, geology, orbital dynamics, astronomy, oceanography, and materials processing. When they are not in the classroom or lugging bags stuffed with heavy manuals, astronaut candidates, or "ascans", receive training in land and sea survival, scuba diving, and being indoctrinated into the use of their spacesuit. Just like you, NASA's trainee astronauts learn to deal with emergencies in the hypobaric environment and they also have the opportunity to spend time experiencing weightlessness in NASA's modified C-9 aircraft, during their parabolic flight training. After a year of basic training, the ascans begin formal space transportation system training, which means they have to spend even more time reading large manuals, taking computer-based training lessons, and becoming proficient in understanding the systems on their vehicle, which, until 2010, will be the Space Shuttle.

Single Systems Trainer (SST) training is the stage that signifies the light at the end of the tunnel for ascans. During this phase of training, an instructor helps the ascan become familiar with the Space Shuttle's systems and subsystems, using checklists that contain information detailing normal system operations and the corrective action to be taken in the event of a malfunction. Following SST training, ascans move on to the more complex Shuttle Mission Simulators (SMSs) that are high-fidelity simulators designed to provide training in all areas of vehicle operations such as launch, ascent, orbital operations, entry, and landing. In parallel with SMS training, ascans spend several training hours in part-task trainers, the Sonny Carter Training Facility that provides controlled neutral buoyancy, and various mockups and trainers such as the payload bay. After SMS training, astronauts wait to receive what they have all been praying for: a flight assignment! By the time an astronaut finally reaches this stage, he/she has accumulated more than 1,000 hours of training. Fortunately for suborbital spaceflight participants, the necessary training has been condensed to a more manageable 21 hours.

Spaceflight theory: Academic Module SA1

The means by which you will reach space is the rocket engine (Figure 3.1, see also color section) mounted at the rear of your vehicle. A rocket engine is simply a jet propulsion device propelled by the expulsion of gases, or propellants, generated in a combustion chamber. The propellants contain both fuel and oxidiser, which means that the rocket engine generates thrust independent of its surrounding environment, unlike the conventional jet engine, which needs to use oxygen from the atmosphere in order to burn fuel.

The thrust that propels the spacecraft is based on Newton's third law of motion, which states that for every action, there is an equal and opposite reaction. To understand how a rocket engine works it is best to think of an analogy such as a closed container filled with compressed gas. Inside this container the trapped gas exerts equal pressure on all surfaces, but if a hole is punched in the base of the container the gas will escape and the pressure within the container will no longer be equal. Because of this pressure differential, the internal gas pressure will push the container upwards

Figure 3.1. The Merlin rocket engine, developed by Elon Musk's SpaceX company (see also color section). Image courtesy: SpaceX. Source: *www.spacex.com/falcon1.php*

in reaction to the jet of air escaping from the base. The gas pressure escaping from the base of this container is thrust, which is determined by the velocity with which burning gases leave the container, or combustion chamber, and the mass of the burning gas, or rocket propellant. In a real rocket engine, the hole in the base of the container is a high expansion ratio nozzle, which is usually bell-shaped, giving the rocket engine its characteristic shape, and it is through this nozzle that thermal energy is converted into kinetic energy, which accelerates the rocket.

Although every vehicle designed to travel into space uses rocket engines, there are some differences. The two primary classes of rocket engine are the solid propellant rocket such as the solid rocket boosters (SRBs) attached to the Space Shuttle, and liquid propellant rockets such as the Saturn V. The SRB, which uses ammonium perchlorate as a fuel, is the largest solid propellant rocket ever built, weighing in at more than half a million kilograms. In contrast, the Saturn V, the largest liquid propellant rocket ever built, used liquid hydrogen as a fuel, and weighed more than 3 million kilograms!

Another class of rocket engine is the hybrid rocket, which uses fuel that is solid and an oxidizer that is liquid (liquid oxygen or nitric acid), carried in a pressurized container above the fuel. One of the problems spaceflight participants may notice when flying onboard a vehicle fitted with a hybrid rocket is the phenomenon of "chugging", which results in significant vibration inside the cabin. Chugging is caused by combustion instability, which is a fairly normal occurrence in hybrid engines and occurs when the liquid nitrous oxide is depleted, leaving only gas.

Another propulsion difference you may notice between current vehicle designs is the utilization of rocket engines *and* jet engines, a feature adopted by EADS Astrium and Rocketplane. In the dual-engine design, the jet engines provide the thrust from takeoff to high altitude, whereupon the rocket engines take over for the final climb into space. In the next spaceflight theory module on Day 4 the mechanics of this climb into space will be explained.

Vehicle indoctrination session: Academic Module VA1

This training module provides you with an overview of the primary systems onboard the vehicle such as the Environmental Control and Life Support System (ECLSS), Flight Control System (FCS), and Thermal Protection System (TPS).

Environmental Control and Life Support System

The ECLSS performs a number of vital functions such as supplying air, maintaining temperature and pressure, as well as shielding passengers from harmful external influences such as radiation. This section provides you with a brief overview of the various systems that comprise the ECLSS and the functions each system provides.

The Air Revitalization System (ARS) is responsible for myriad functions, one of which is ensuring humidity remains between 30% and 75%. The ARS also ensures carbon dioxide and carbon monoxide levels remain non-toxic, that temperature and ventilation is regulated, and that the vehicle's avionics and electronics are cooled. To achieve this, the ARS consists of a series of water coolant loops, air loops, and heat exchangers, the layout of which will be specific to your vehicle.

The Active Thermal Control System (ATCS) component of the ECLSS is responsible for heat rejection, a function it achieves by the use of cold-plate networks, coolant loops, liquid heat exchangers, and various other heat sink systems that reject heat outside the vehicle. The vehicle has a number of systems that generate heat and it is important that the heat sink systems are not overloaded, although if the capacity of the heat sink unit is exceeded the vehicle may be able to activate a flash evaporator that is designed to meet excess heat rejection requirements for short periods.

Each spacecraft has a Guidance, Navigation, and Control (GNC) system, which is designed to control the attitude and position of the vehicle during its mission. As the acronym suggests, GNC comprises the sub-areas of guidance, navigation, and control. Everyone is familiar with navigation, which, when applied to spacecraft, involves computing the orientation and position of the spacecraft with reference to a rotating reference system such as the Earth. There are several local (onboard your vehicle) and remote sensors (on the ground) which collect data, which is then processed by onboard and terrestrially located computers. Once this information has been interpreted, the navigation information is relayed to the pilot informing him where he is. Although this information is useful it is also advantageous to know where the spacecraft will be at some future point in time and it is here that the principle of guidance is helpful.

Guidance is able to predict the future position of the spacecraft and compare this information with the intended profile that is stored in the flight computer. Guidance works by propagating information such as environmental torques and forces acting on the spacecraft, and by means of complex mathematical algorithms uses this information to predict what the spacecraft will do next.

Working in conjunction with navigation and guidance is the process of control, which ensures the spacecraft is oriented and moving in the intended direction, directed by guidance. Some aspects of control are performed automatically by engaging attitude maneuvrring thrusters and reaction control jets to ensure the spacecraft is moving along the intended trajectory whereas the Flight Control System (FCS) directs other aspects.

Flight Control System

Your vehicle's flight controls are similar to a modern airliner, which use digital fly-by-wire control systems run by a computer to operate its flight surfaces. The Flight Control System (FCS) consists of the flight control surfaces, the cockpit controls, and the mechanisms responsible for controlling the vehicle in flight. To avoid the problem of a computer crashing, your vehicle's FCS is triply redundant, meaning that it has three computers in parallel, and three separate wires leading to each control surface, so that even if two computers crash there will still be one capable of flying the vehicle.

Reaction Control System

To enable the vehicle to maneuver exoatmospherically it is fitted with a Reaction Control System (RCS) that consists of several cold-nitrogen gas thrusters placed in opposing pairs so they can change the attitude of the vehicle in all three dimensions. By making small thrust alterations the pilot can orientate the vehicle to ensure not only that you will have the best views of Earth and space, but also to realign the vehicle prior to re-entry.

Thermal Protection System

If you have ever watched a Space Shuttle mission you will be familiar with thermal protection terms such as "heat shields" and "shuttle tiles". The goal of a thermal protection system (TPS) is to prevent excessive heat from damaging or destroying the vehicle and its contents, which was the fate of the Space Shuttle Columbia during her re-entry to the Earth's atmosphere in 2003. The problem for spacecraft designers is to keep the thermal protection materials to a minimum weight, as there are several surfaces on the vehicle that require protection from atmospheric heating during re-entry. The wings, tail, flight control surfaces, and nosecone of the vehicle each experience high temperatures and must be protected by some type of temperature-resistant material, usually titanium coated with a special ceramic paint.

Data Acquisition System and Integrated Vehicle Health Management System

The Data Acquisition System (DAS) collects data from hundreds of sensors located throughout the vehicle. These sensors transmit information to Mission Control and to the Integrated Vehicle Health Management System (IVHMS), which uses the data to check for anomalous readings from the vehicle's systems and structure. If the IVHMS detects any off-nominal event, it will attempt to diagnose the problem, and if it deems the anomaly to be of concern it will warn the pilot.

High-altitude indoctrination training, Part I: Academic Module HA1

Anyone who has watched such movies as *Outland* and *Sunshine* will know the space environment is highly dangerous without proper protection. In both these movies, characters are killed by an explosive decompression, and it is such an event that is perhaps the greatest threat to you during your suborbital flight. However, despite the graphic depiction of exploding bodies portrayed in *Outland*, the body, thanks to the containing tension of the skin, will not explode. In fact, it is possible to be exposed to a complete vacuum for up to 30 seconds and suffer no permanent physical damage as depicted in *Sunshine*, although if you were to spend any longer the chances of you surviving would be remote. Few humans have been unfortunate enough to test this theory, but incidents have occurred in which people have been exposed to near-vacuum conditions on Earth, as described in the following NASA report from 1965.

"At NASA's Manned Spacecraft Center we had a test subject accidentally exposed to a near vacuum in an incident involving a leaking spacesuit in a vacuum chamber back in '65. He remained conscious for about 14 seconds, which is about the time it takes for oxygen deprived blood to go from the lungs to the brain. The suit probably did not reach a hard vacuum, and we began repressurizing the chamber within 15 seconds. The subject regained consciousness at around 15,000 feet equivalent altitude. The subject later reported that he could feel and hear the air leaking out, and his last conscious memory was of water on his tongue beginning to boil." [5]

The risk of decompression is increased once you exceed the "Armstrong Line" at an altitude of 19,090 m, since at this pressure water will boil at the normal temperature of the body ($37°C$). Above the altitude of the Armstrong Line, exposed human fluids will boil without the protection of a pressure suit and no amount of oxygen will sustain life for more than a few minutes.

If, in the worst case scenario, your vehicle suffers an explosive decompression and you happen to be flying with an operator that does not require its passengers to wear pressure suits you can expect to experience several life-threatening effects. First, due to the tremendous pressure differential between your lungs and the near vacuum of space, your lungs will tear and quickly rupture. Second, your heart rate will shoot up and then fall and you will feel pressure in your chest as your heart stretches due to absence of oxygen (anoxia). Third, you will notice your skin start to swell and then rapidly distend, so that eventually your body may be almost twice its size. As Tamarack Czarnik states in his classic work entitled *Surviving Rapid/Explosive Decompression* [2], "Our patient will look no better than he feels, though this means little in terms of survival." If you are wearing a pressure suit such as the Russian

Sokol suit that Starchaser will require its passengers to wear you will be protected but if the suit is a partial "get me home" suit you will be afforded only protection for about 30 minutes, but by this time the vehicle should have descended to a life-sustaining altitude.

The high-altitude indoctrination (HAI) course is designed to help you recognize early the symptoms and signs of subtle effects of hypoxia and to teach you the correct use of the oxygen mask and regulator. It will also teach you the Valsalva maneuver for clearing your ears, demonstrate the effects for rapid decompression, and teach you pressure breathing techniques under controlled conditions. In addition, this training will identify any Eustachian tube dysfunction and will teach you communication while wearing a mask. It will also identify any claustrophobia-prone individuals.

AFTERNOON OF DAY 2

Basic space physiology: Academic Module PA1

One of the most profound changes you will experience during your short journey into space is the absence of the Earth's gravitational pull. In microgravity, your body's automatic reaction to maintain posture will be absent, since the pressure receptors in your feet and ankles will no longer signal the direction of "down". This sensation will be compounded by the fact that the otolith organs located in your middle ear, which are sensitive to linear accelerations, will also no longer perceive a *down* direction. This absence of sensation will cause you to become disoriented and susceptible to visual orientation illusions, which in turn may cause you to feel ill. You will experience similar symptoms during your exposure to short periods of microgravity onboard the 727-200 aircraft during your zero-G indoctrination.

Initially, you will probably experience an increase in heart rate due to the anticipation of what is about to happen. For example, typical pulse rates recorded prior to launch are in the 160 to 180 beats per minute range, a heart rate which will usually be accompanied by an increase in respiration rate from 12 to 24 breaths per minute. Body temperature may increase slightly during the flight, but these changes will be insignificant. Your heart rate and respiration rate will decline during deceleration prior to achieving suborbit, but your heart rate will increase again during re-entry and you may also experience a tumbling sensation as you coast to suborbit [2].

A second effect that you will feel toward the end of your three to five minutes of microgravity is the effects of fluid being shifted inside your body due to the absence of gravity.

Survival training: Practical Module SP1

This module will familiarize you with the contents of the vehicle's survival equipment, which contains the items listed below, the use of which will be described to you by your instructors during your survival training phase.

a. Utility knife
b. Waterproof and windproof matches (24)
c. PDA containing topographic world maps
d. Aviator™ flashlight (1) and lithium batteries (4)
e. Binoculars (1)
f. Emergency freeze-dried foods (15 meals) and food-gathering equipment
g. Signaling items: flares (5), mirror (1), electronic beacon (1), xenon strobe (1)
h. Parachute flares (3)
i. 150 ft of paracord.
j. Miox™ disinfection pen (1) and batteries (4)
k. Magellan handheld GPS unit.

Since a segment of your flight occurs over open ocean, your vehicle also contains the following life raft equipment:

a. Life raft
b. Bailer
c. Graduated drinking vessel
d. Pump, leak stoppers, and raft repair kit
e. Anti-seasickness medication
f. Buoyant paddles
g. Sea anchor
h. Radar reflector
i. Buoyant orange smoke
j. Thermal protective aids (3)
k. Non-thirst-provoking rations (24 hours).

High-altitude indoctrination training, Part II: Academic Module HA2

In the Physiological Zone, which extends from sea level to 3,000 m no oxygen or special protective equipment is required. The Physiologically Deficient Zone, which extends from 3,000 m to 15,000 m, is characterized by reduced pressure and oxygen deficiency and requires you to use supplemental oxygen since, if your vehicle undergoes a rapid decompression, you will be exposed instantly to a lower air density or pressure. For example, at 15,000 m the air pressure is less than a quarter of what it is at sea level. This reduction in air pressure will impose a number of possible adverse effects on you, one of which is gas expansion within your body cavities. Before we examine these effects it is appropriate to review some of the gas laws.

If you can remember your physics classes you may recall that as pressure falls a given amount of gas will expand as long as temperature and mass remain constant (Boyle's Law). There are a number of gas cavities within your body and if you are exposed to a high altitude suddenly these cavities will expand, causing all sorts of problems such as trapped air in a cavity such as a tooth, which will cause an excruciating condition known as barodontalgia. Other gas-containing cavities include the lungs, the air passages, the stomach, and the middle ear cavity. The latter separates the middle ear cavity from the outside, which, if subjected to a sudden drop in outside pressure, may rupture or perforate, which can be quite a disabling occurrence.

Another problem caused by falling pressure is hypoxia, which is an acute syndrome caused by oxygen deficiency. If not treated, hypoxia may rapidly aggravate to anoxia, which is an absence of oxygen. If your vehicle suffers a rapid decompression you will experience the signs and symptoms of hypoxia that include increased respiration, cyanosis, mental confusion, hallucination, memory loss, poor judgment, blurred vision, tingling, myclonic jerks, and eventually unconsciousness. The length of time an individual is able to perform useful activity in such an event is known as time of useful consciousness (TUC) and is a measure of the period of time from the interruption of the oxygen supply or exposure to an oxygen-poor environment, to the time when useful function is lost. At high altitudes, or in the event of a rapid decompression in which oxygen is sucked out of the cabin, TUC becomes very short and it is therefore important that passengers be able to recognize the symptoms of oxygen deprivation.

G-tolerance training: Academic Module GA1

The theoretical component of this training will take place at your Operator Training Facility and the practical phase will be conducted at the NASTAR training facility in Southampton, Pennsylvania.

During your suborbital flight you will encounter different acceleration stresses and although the vehicle has been designed to reduce this stress in the case of malfunction, especially during re-entry, these forces may be very large. It is necessary therefore, that you receive training in G tolerance. This section provides you with an introduction to the theoretical and practical aspects of G tolerance and the accelerative forces you will encounter during launch and re-entry.

Sustained acceleration, usually written as $+G_z$, is acceleration that lasts for more than 1 second and is a force that has the potential to crush your windpipe and rupture your air sacs, which will make it impossible for you to breathe. High rates of sustained acceleration can also result in blood pooling to such a degree that it will cause you to convulse and eventually black out. Given the serious consequences of these events it is important you understand the effects so you are able to deal with inflight events such as grayout, blackout, or even unconsciousness.

During your suborbital trip to you can expect to experience four distinct phases of accelerative stress, each differing in their magnitude and duration.

a. *Launch*. Between 1.5 G and 4.0 G depending on the mission architecture of your operator.
b. *Suborbital*. During this phase the centrifugal force of the spacecraft balances the gravitational force, thus producing a microgravity environment!
c. *Re-entry*. You will begin to notice acceleration stresses shortly after the end of the microgravity segment of your flight, but these should be no more than 1.5 G to 3 G.

d. *Emergency egress.* The forces experienced during an emergency egress will be
    different in different phases of flight, but you can expect high-magnitude
    accelerations that exceed 15 G, sustained for 1 or 2 seconds.

Cardiovascular effects of $+G_z$

Because the cardiovascular system is the most sensitive of the physiological systems
to $+G_z$ you will be instrumented with ECG and heart rate–monitoring equipment
before your run so that you will be able to see for yourself how your body reacts to
increasing G. Generally you can expect your heart rate to correlate with increased
$+G_z$ due to the acceleration force effect and the general psychophysiological stress
syndrome that is associated with exposure to acceleration.

   Many of the central nervous system (CNS) effects of $+G_z$ are a direct consequence of cardiovascular effects, since a regular blood supply is required for the CNS
to function, so the ability of your body to tolerate acceleration is related directly to
adequate blood flowing to your brain. Because of this relationship, symptoms that
relate to insufficient blood flow to the brain are used to determine tolerance to $+G_z$.
The normal index of defining G-level tolerance is to use loss of vision in an upright-
seated position at a specific level of G exposure, as outlined in Table 3.2.

Table 3.2. Categorization of light-loss criteria.

Symptom	*Description*	*Onset of symptoms*	*Criteria*
Grayout	Partial LOV. Often occurs as the first physiological effect of sustained G loads. Low blood oxygen levels cause peripheral vision to fade. Objects in the center of the FOV can be seen but appear to be surrounded by gray haze	3.5 G	100% peripheral light loss (PLL) combined with 50% central light loss (CLL)
Blackout	Gray haze envelops entire FOV and almost immediately becomes black. You will be conscious but unable to see	Above 5 G	100% CLL, but sufficient blood reaches brain to permit consciousness and hearing
Gravity-induced loss of consciousness (G-LOC)	Follows quickly after blackout with sustained G load. You will be unconscious but will regain consciousness when G load is released	Above 5 G	Normally occurs following increase of acceleration after blackout

LOV = loss of vision. FOV = field of vision.

Individual tolerance to $+G_z$

Tolerance to $+G_z$ may vary from day to day and is highly individualized due to differences in physiological responses. You should ensure that you have eaten prior to the centrifuge run because if you are hypoglycemic you will impair your heart's ability to compensate at the onset of high-G loads, and will experience grayout and blackout at relatively low sustained G loads. If you are unfit you can expect a significant decrease in your ability to tolerate $+G_z$. If you have an illness you should inform the staff at the NASTAR center, as most illnesses will compromise your tolerance to $+G_z$. Also, be sure to drink adequately prior to your run, as dehydration will reduce your G tolerance by reducing plasma volume.

DAY 3

Time	Training module	Details	Location
08:00–08:50	Review of high-altitude indoctrination	Theory. Required reading: Section 4 of *SFP Handbook*	NASTAR
09:00–09:50	*CP1* Hyperbaric chamber familiarisation	Practical. Required reading: Section 4 of *SFP Handbook*	NASTAR
10:00–10:20	Coffee break		
10:20–11:10	*CP2* Commencement of 30 min prebreathe	Practical. Required reading: Section 4 of *SFP Handbook*	NASTAR
11:20–12:10	*CP3* Chamber run to 7,500 m. Rapid decompression	Practical. Required reading: Section 4 of *SFP Handbook*	NASTAR
12:15–13:05	Lunch break		
13:10–14:00	*GA2* Review of G-tolerance training. Centrifuge orientation	Theory. Required reading: Section 3 of *SFP Handbook*	NASTAR
14:10–15:00	*GP1* Centrifuge runs 1–5 for crewmembers 1 and 2	Practical. Required reading: Section 4 of *SFP Handbook*	NASTAR
15:05–15:25	Coffee break		
15:30–16:20	*GP2* Centrifuge runs 1–5 for crewmembers 3 and 4	Practical. Required reading: Section 4 of *SFP handbook*	NASTAR
16:30–17:20	Review. Presentation of HAI and G certification	Practical. Required reading: Section 4 of *SFP Handbook*	NASTAR
17:30–18:20	Dinner		
18:30–20:30	Fly to operator's training facility		
21:00–22:00	Check back into your hotel		

MORNING OF DAY 3

Hypobaric chamber orientation: Practical Module HP1

Your HAI practical phase will be conducted at the National Aerospace Training and Research Center (NASTAR), located in Bucks County, Pennsylvania.

When you enter the hypobaric chamber (Figure 3.2) you will be shown to your seats and consoles. The consoles are located between your seats and contain an intercom cord and an oxygen pressure gauge with an *on* and *off* control. The instructor will then orientate you to the function and layout of the chamber and explain the purpose of the observer station, which is supplied with life support equipment, including oxygen control, breathing hoses ands fittings, pressure gauges, and an intercom station with channel mode selector switches. The hoses are long enough to allow the instructor to reach any crewmember within the chamber. The chamber itself is capable of being evacuated to an altitude-equivalent atmosphere of 303,030 m at a rate of up to 1,515 m per minute.

AFTERNOON OF DAY 3

Centrifuge training: Module GP1

On arrival at NASTAR you will review the major theory elements of Module GA1 and be taken on a tour of the facilities.

The NASTAR centrifuge (Figure 3.3, see also color section) has the capability of a G-onset rate of 6 G per second compared with the 0.1 G/s you will experience during your run. During your indoctrination to the facility you will be shown the interior of the gondola and your instructor will point out the adjustable rudder pedals provided for foot support, and the shoulder and lap harnesses that will secure you. At this point you will have the opportunity to try on the facemask you will be wearing during the run to monitor your breathing. The facemask serves a dual function as it also allows two-way communication with the console operator. When you sit in the chair you will probably notice a small video camera, which will record you during the run.

Shortly after breakfast on Day 2 you will have the opportunity to observe a dry run from the NASTAR console room. The console operator will review the G-onset loads and run through the operation of the communication system and will then assign you to a centrifuge rotation and explain the safety procedures and the role of the flight surgeon and centrifuge operator. Finally, before you step into the gondola the flight surgeon will explain what you should expect during each run. Support personnel will then take over and supervise your ingress into the gondola where you will be connected to biomedical instrumentation that will include a 12-lead ECG, blood pressure cuffs, and respiratory monitoring equipment.

The run schedule for your initial assessment is detailed in Table 3.3.

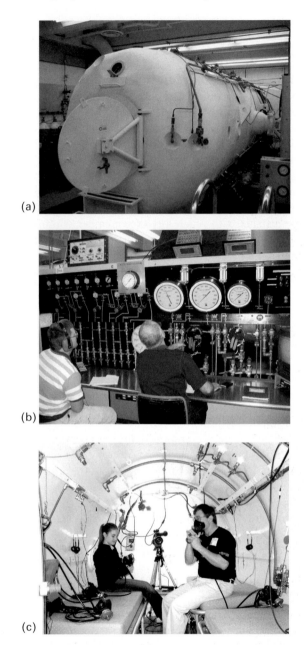

(a)

(b)

(c)

Figure 3.2. (a) The high-altitude indoctrination component of your training will take place inside a hypobaric chamber similar to the one depicted above. Image courtesy: Simon Fraser University. (b) The control panel of the hyperbaric chamber in which spaceflight participants will receive their high-altitude indoctrination training. Image courtesy: Duncan Milne and Simon Fraser University. (c) The interior of Simon Fraser University's six-man hyperbaric chamber. Image courtesy: Duncan Milne and Simon Fraser University.

(a)

(b)

Figure 3.3. The G training will take place at NASTAR's environmental training facility, located in Southampton, Pennsylvania. (a) The centrifuge in static mode (see also color section). (b) The control room from which you will be monitored. Image courtesy: Environmental Tectonics Corporation. Source: *www.etctacticalflight.com/atfs_400gallery.php*

Table 3.3. Run schedule for determination of G sensitivity.

Run #	Type of run	Rate of onset	Peak G		Rate of offset
		(G/s)	Magnitude	Duration (s)	(G/s)
1	Warm-up	0.5	2.5	15	0.2
2	GOR	0.1	5.0	5	1.0
3	ROR 1	1.0	3.0	10	1.0
4	ROR 2	1.0	4.0	15	1.0
5	ROR 3	1.0	5.0	20	1.0

GOR = gradual onset run. ROR = rapid onset run.

DAY 4

Time	Training module	Details	Location
08:00-08:50	*ZA1* Zero-G theory. G-Force-One orientation	Theory. Required reading: Section 5 of *SFP Handbook*	Operator Training Facility
09:00–11:00	*ZP2* Zero-G training. 15 parabolas	Practical. Required reading: Section 5 of *SFP Handbook*	Operator Training Facility.
11:20–12:10	*SA1* Spaceflight theory. Suborbital trajectory	Theory. PPT. Required reading: Section 2 of *SFP Handbook*	Operator Training Facility.
12:15–13:05	Lunch break		
13:10–14:00	*LE1* Launch Escape System	Theory. PPT. Required reading: *Operator Vehicle Handbook*	Operator Training Facility
14:10–15:00	*LE2* Launch Escape System training	Practical	Operator Training Facility
15:05–15:25	Coffee break		
15:30–16:20	*SP1* Survival training	Practical. Required reading: Section 4 of *SFP Handbook*	Operator Training Facility
16:30–17:20	*EP1* Vehicle simulator.	Practical. Required reading: *Operator Simulator Handbook*	Operator Training Facility
17:30–18:20	Dinner and *OP1*: Meet the flight crew and ground teams		
18:30–20:30	Personal administration and preparation for next day's suborbital flight		

MORNING OF DAY 4

Basics of zero-G: Academic Module ZA1

Parabolic flight, also known as zero-G flight, is achieved by flying a specially modified aircraft through a parabolic flight maneuver between altitudes of 7,500 m and 10,500 m. Each parabola takes ten miles of airspace to perform and lasts approximately one minute from the pull-up to pull-out phase. A typical parabolic flight similar to the one you will be experiencing will involve 15 parabolas, providing you with a cumulative microgravity exposure that will be longer (about 6 minutes) than you will experience during your actual suborbital flight!

As you can see in Figure 3.4 the parabolic flight maneuver is a little like a rollercoaster maneuver that requires the aircraft to pull up in a "nose-high" 45° angle of attack. During this phase of the flight you will begin to feel heavy, as the G loading will be nearly two times normal Earth gravity. As the aircraft approaches the zero-G segment of the flight it will "push over' the top, a portion of the flight that is usually referred to as insertion. At this point, you and everything that is not fixed inside the cabin, will start to float. For the next euphoric 25 to 30 seconds, depending on the skill of the pilot, you will experience the sensations of microgravity, sometimes called weightlessness, although technically the type of weightlessness that occurs during a parabolic flight is actually more related to freefall. At the end of the freefall period you will be alerted that the pull-out phase is about to begin, which occurs as the aircraft pushes its nose about 30° down to allow all the passengers to secure themselves on the floor of the aircraft. Gradually the G force builds up again to 1.8 until the aircraft has regained its initial altitude of 7,500 m.

After a short recovery period, during which you and some of your fellow passengers may have been sick, the whole process starts again, to be repeated another 14 times.

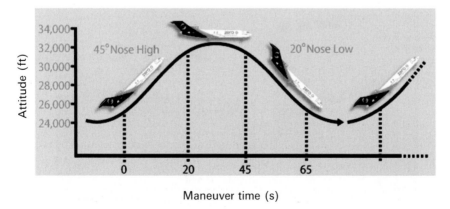

Figure 3.4. Parabolic flight maneuver. Image courtesy: Go Zero G. Source: *www.gozerog.com/how-it-works.htm*

Zero-G: Practical Module ZP1

You will experience your first taste of weightlessness onboard an aircraft named G-Force-One (Figure 3.5), a specially modified 727-200 that belongs to the Zero Gravity Corporation, a privately held space entertainment and tourism company located in Las Vegas. The aircraft is able to accommodate up to 35 passengers and 6 crewmembers, and is licensed in accordance with the FAA's Aircraft Certification Office requirements. Interestingly, NASA operates a C-9 aircraft, which it uses for research and training of its astronauts, but because the FAA does not certify the aircraft, NASA is not allowed to use it for flying the public.

Your parabolic flight will begin like any commercial flight, by taxiing onto a runway for a takeoff clearance. Following takeoff the aircraft will fly to the parabolic flight FAA-designated airspace, approximately 160 km long and 16 km wide, where the parabolic flights will be conducted. Shortly before the first of the parabolas, your instructor will tell you to lie on the padded floor, designated as a floating area. While you lie there you may be wondering about the safety of what you are about to do, but you needn't worry. Zero-G flights have an outstanding safety record stretching back 45 years that includes flights conducted by the Russian Space Agency, the European Space Agency, and, of course, NASA, which has conducted more than 150,000 parabolas on aircraft such as the KC-135 and the Learjet-25. During the actual parabola you are free to do what you like, whether this involves attempting Olympic gymnastics, playing with a handful of M&Ms, or just flying up and down the cabin.

The takeoff, parabolic flight segment, and landing will last between 90 and 100 minutes, after which you have the option of celebrating with champagne or considering if your stomach will be able to handle the real deal! Having said that, it is unlikely

Figure 3.5. G-Force-One, a modified 727-200. Image courtesy Go Zero G. Source: *www.flickr.com/photos/jurvetson/214792903/in/set-5957/*

you will suffer motion sickness as you will fly only 15 parabolas, which is below the number (typically 25 to 30 parabolas) of parabolas that normally induces motion sickness.

Spaceflight theory: suborbital trajectory and flight dynamics: Academic Module SA2

The vehicle that will take you into space is designed specifically for suborbital flight, which means that the spacecraft will reach space but will not enter a stable orbit. A suborbital flight is different from a flight that reaches orbit but then deorbits after less than one complete orbit of the Earth. Such a flight, which is the same as Yuri Gagarin's flight onboard Vostok 1 on April 12th, 1961, is considered a flight to LEO.

To achieve the desired altitude of 100 km and allow passengers to experience a few minutes of microgravity, it is necessary for the vehicle's minimum velocity to be 1.4 km/s, which compares with the velocity of 10 km/s that is required to reach LEO of approximately 300 km. Another component in the equation is the horizontal distance that the spacecraft will travel during its flight, since the greater the horizontal distance the greater the required velocity. Regardless of how much velocity is required, any suborbital vehicle that reaches space must undergo atmospheric re-entry and the aerodynamic heating that occurs during this process. The degree of aerodynamic heating is determined by the maximum speed of the flight, so that a flight with a maximum velocity of 1.5 km/s will experience less heating than a flight with a maximum velocity of 7 km/s.

The three takeoff modes available to commercial launch operators include vertical, which has been adopted by Starchaser; horizontal, which is Virgin Galactic's choice; and air launch, which is incorporated into the da Vinci Project's mission architecture. The landing modes available include wings/parafoil which is considered a horizontal landing mode and aerodynamic decelerators such as parachutes, rockets, and rotors, which are considered vertical landing modes. Of these landing methods, only vertical landing has not been demonstrated as a landing method for manned Earth-return missions, although many tests have been conducted using rockets only.

Ascent trajectory

In order to achieve the desired altitude it is necessary for operators to use a rocket engine that will provide the vehicle with sufficient thrust, a quantity that is called specific impulse and abbreviated "I_{sp}". The specific impulse of a rocket engine is defined as *the number of seconds a kilogram of propellant will produce a kilogram of thrust* and is a measure that is analogous to "kilometers per litre" of a car engine. For example, a specific impulse of 250 seconds means that a rocket engine would consume one kilogram of propellant when producing one kilogram of thrust for 250 seconds. Although rocket engineers try to design an engine that has the highest I_{sp}, the designs of the current suborbital vehicles generally have specific impulses between 200 and 275 seconds. Once the I_{sp} is known, the maximum velocity can be calculated from a load of propellant, using Tsiolkovsky's equation [7], an explanation of which is beyond the scope of this discussion.

Once I_{sp} and propellant factors have been calculated, rocket engineers must then determine delta-v, which is provided by the thrust of a rocket engine, whereas the time–rate change of delta-v is the magnitude of the acceleration caused by the engines and is therefore a quantity that is directly related to the amount of fuel that the vehicle must carry, which in turn influences the trajectory of the vehicle. The available delta-v is a direct function of the efficiency of the fuel being used, which means that the higher the energy available, the higher the resultant delta-v.

The delta-v calculation is a complex one since it must take into consideration the losses imposed as a vehicle climbs into space. The first loss is gravity loss which is caused by the vehicle's engine trying to pull away from the Earth's gravity well. The second loss is drag, which arises as a result of the friction between the vehicle and the atmosphere. Since most drag losses occur in the supersonic flight regime, rocket engineers try to design a rocket that is long and sleek. Another way of avoiding drag is air launch since launching above much of the atmosphere significantly reduces drag. The third loss is the need to steer the vehicle, which is achieved by using aerodynamic control surfaces. To minimize the effects of steering, rocket engineers spend hundreds of hours conducting flight trajectory simulation computer programs.

Once the rocket engineer has calculated all the losses, he/she must determine other factors that impact delta-v such as the residual and reserve propellant loads. Residual propellant is the fuel trapped in the propellant tanks and lines, whereas the reserve propellant is the extra propellant loaded on the vehicle so there is sufficient contingency fuel for the flight. This information is then combined with the takeoff mass of the vehicle which is the sum of the crew, passengers, propellant mass, and the empty mass of the vehicle itself. Once all these factors have been calculated the delta-v of the vehicle can be determined and the numbers can be crunched to see if the engines have sufficient power to enable the vehicle to achieve the desired altitude.

Launch modes

Vertical takeoff. Starchaser Industries' Thunderstar vehicle is an example of a vertical takeoff vehicle that launches from a gantry takeoff pad in a vertical orientation very similar to the Soyuz rocket that takes cosmonauts to the ISS. Incidentally, due to the little launch preparation time and safety advantages offered by vertical takeoff, it is this mode that has been selected by every space transportation study in the past 20 years as the preferred takeoff mode [6].

Horizontal takeoff. XCOR's Xerus spacecraft and Rocketplane's DreamChaser™ are both examples of vehicles that utilize horizontal takeoff. The vehicle designs of each company feature a winged craft with much of the fuselage occupied by propellant tanks and both jet and rocket engines, the design of which has both advantages and disadvantages. First, wings reduce the delta-v required to reach suborbital altitude, with most of the benefit occurring at high altitude, but this is negated to a degree by the additional drag caused by the wings themselves, which of

views through the windows and every switch and toggle is exactly the same as in the real vehicle. Computer-generated sound simulations come from hidden speakers, which duplicate those sounds you will hear during your flight, including the aerodynamic vibrations, thruster firings, rocket ignition, and landing gear deployment. Your pilot will have spent many hours training in this simulator, dealing with all the possible failures and problems that may occur. Your session in the simulator will take you from $T - 5$ minutes and through the phases of launch, ascent, suborbit, re-entry, approach, and landing.

DAY 5

Time	Training module	Details	Location
08:00–08:50	Breakfast with fellow crewmembers		Operator Training Facility
09:00–09:50	Preflight check	Suit check with technicians. Egress into vehicle	Operator Training Facility
10:00–12:05	**SUBORBITAL FLIGHT!**		
12:15–13:05	Lunch break		
13:10–14:00	Astronaut Wings ceremony	Presentation of FAA civilian astronaut wings	Operator Training Facility
14:10–15:00	Debrief and return of training equipment		Operator Training Facility
15:05-15:25	Coffee break		
15:30–16:20	Return to airport		
16:30–17:20	Return home		

Due to the different mission architectures employed by space tourism companies, the following is a generic account of what you can expect during your suborbital flight.

MORNING OF DAY 5

It is 7 o'clock on Day 5 and it is time to begin the final preparations for the flight of your life. You have already been awake for two hours in anticipation of this day and since you slept in your spacesuit you don't have to worry about getting changed!

You check yourself out in the mirror for the 50th time, paying particular attention to the mission patch sown onto the left arm of the suit that reminds you that this time it is for real. You rummage through your personal flight case and check again you have everything. Mini-DVD camera? Check. Digital still camera? Check. Mission pins? Check. The ALF mascot your daughter wanted you to take up? Check. You've waited a long time, spent a lot of money, and invested in a lot of training for this day to become reality, but today is the day that will change your life and your perception of Earth. You make your way with family and friends to the spaceport restaurant where, after a routine security briefing, you eat a breakfast together with the other candidate spaceflight participants. After a photo shoot and a final check of your spacesuit, you enter the doors of the MCC where you say your final goodbyes to family and friends. They will stay here and watch on the monitors. You give a final wave and then board the spacecraft where, with the assistance of the technicians, you settle into the seats as you listen to the ex–Space Shuttle pilot brief you on the flight. The technicians give you a final check, ensuring you have fastened your five-point harness and then, with a thumbs-up, they leave the vehicle.

It is now just you, three spaceflight participants, and the pilot. After receiving taxi clearance from the spaceport traffic control, the spacecraft taxis onto the runway and with a kick of the jet engines, lifts its wheels off the runway, taking off just like a business jet that it closely resembles. After a leisurely 45-minute cruise you reach an altitude of 12 km and the pilot briefs you to prepare for rocket ignition. Moments later, with a flick of the pilot's switch, the rocket engine is ignited and you feel like you've been punched in the back. As the G forces build inexorably, you are pushed further and further back into your seat. Within seconds the vehicle is climbing almost vertically as it accelerates to Mach 1. You look out of the window and you notice the blue sky becoming noticeably darker with each passing second. Less than a minute after rocket ignition the pilot informs you that you are now traveling faster than Mach 3 and there are fewer than ten seconds before completion of engine burn. At 60,000 m you hear the pilot announce that he is switching off the engines and a moment later the cabin falls silent. Your view through the window is nothing short of spectacular, a view that less than a thousand before you have experienced. Ever. You slowly become aware of the sensations of microgravity, just like your ride in G-Force-One. The vehicle is now more than 100,000 m above the Earth and you have officially earned your spaceflight participant wings, and in doing so, placed yourself in the select group of those humans who can say they have flown in space.

Inevitably, the three minutes of weightlessness passes all too quickly as you hear the pilot request that you take your seats for the beginning of the descent maneuver. You begin to appreciate your G-tolerance training in NASTAR's centrifuge as the G forces build, once again pushing you into your seat so that you weigh almost four times your bodyweight during your rollercoaster ride back to Earth. The friction in the atmosphere gradually slows the vehicle to subsonic speeds as it begins a gradual glide to the runway. You hear the familiar hum of the jet engines as the vehicle flies back to a perfect landing at the spaceport from which you took off less than two hours ago.

As the vehicle taxis onto the apron you can see family and friends waiting to hear about your experience. After greeting them with a big smile, you follow your crewmembers to the reception at the spaceport for the presentation of your FAA civilian astronaut wings.

REFERENCES

[1] *Commercial Space Launch Amendments Act*, H.R. 5382, 2004. Online at *www.faa.gov/about/office_org/headquarters_offices/ast/media/PL108-492.pdf*

[2] Czarnik, T.R. *Ebullism at 1 Million Feet: Surviving Rapid/Explosive Decompression*. Online at *www.sff.net/people/Geoffrey.Landis/ebullism.html*

[3] *FAA Guide for Aviation Medical Examiners*. Online at *www.faa.gov/about/office_org/headquarters_offices/avs/offices/aam/ame/guide/media/guide06.pdf*

[4] Fédération Aéronautique Internationale. *Human Space Flight Requirements for Crew and Space Flight Participants: Final Rule*. Online at *http://edocket.access.gpo.gov/2006/pdf/E6-21193.pdf*

[5] Roth. E.M. *Rapid (Explosive) Decompression Emergencies in Pressure-Suited Subjects*. NASA CR-1223, November 1968.

[6] Sarigul-Klijn, M.; and Sarigul-Klijn, N. *Flight Mechanics of Manned Sub-Orbital Reusable launch Vehicles with Recommendations for Launch and Recovery*. AIAA 2003-0909, pp. 5–6, January 2003.

[7] Tsiolkovsky, K. *Exploration of the Universe with Reaction Machines*. Science Review No. 5, St. Petersburg, 1903.

4

Orbital flight: The orbital experience, company profiles, mission architectures, and enabling technologies

> *For once you have tasted flight you will walk the earth with your eyes turned*
> *skywards, for there you have been and there you long to return.*
>
> Leonardo da Vinci

If a suborbital flight has whetted your appetite for more of the same, or if your trip to suborbit didn't excite you enough, then you may be considering a trip into orbit. To help you choose a company that offers an orbital flight experience, this chapter provides an overview of those commercial space entities planning to offer tickets within the next decade. Also discussed are the risks involved in sending passengers into orbit and the abort modes and emergency egress systems that the prospective orbital spaceflight participant should consider before choosing an operator.

CHANGING THE ORBITAL LAUNCH INDUSTRY

Many of the skyward-looking entrepreneurs involved in the private space business do not claim to have a secret formula for dramatically lowering launch costs, just sound business sense, engineering skills, and the ability to choose and lead the best team money can buy. Unlike Robert T. Bigelow, CEO of Bigelow Aerospace, many have no intentions of becoming an astronaut, but are instead motivated to change the current stagnant marketplace. Each chief executive officer of the nascent orbital space industry is confident there is a market potential for orbital flight and that there is money to be made outside the atmosphere. Although the first batch of $5 million orbital tickets will not be cheap, entreprenauts such as Bigelow and Elon Musk are gambling that there are enough adventurous millionaires to support an orbital spaceflight industry.

A major hurdle facing any non-government space entrepreneur committed to launching passengers into orbit is the prohibitive cost. However, due to NASA's requirement to deliver crew and cargo to the International Space Station (ISS) following the Space Shuttle's retirement in 2010, the U.S. space agency will be reliant on the private sector, at least until 2015, the year when NASA's Orion crew transfer vehicle should become operational. In an attempt to offer financial assistance to private sector companies such as Bigelow Aerospace, Rocketplane Kistler, and SpaceX, NASA, through its Exploration Systems Mission Directorate (ESMD), recently created the Commercial Orbital Transportation Services (COTS), a program designed to help commercial providers build spacecraft that NASA can contract for the purpose of ferrying its astronauts and cargo to ISS. To achieve this, NASA is spending $500 million (less than the cost of a Space Shuttle mission!) to finance demonstration orbital vehicles that are being built by commercial providers.

NASA Administrator Michael Griffin's COTS message states

> "I believe that with the advent of the ISS, there will exist for the first time a strong, identifiable market for 'routine' transportation services to and from LEO, and that this will only be the first step in what will be a huge opportunity for truly commercial space enterprise, inherent to the Vision for Space Exploration. I believe that the ISS provides a tremendous opportunity to promote commercial space ventures that will help us meet our exploration objectives and at the same time create new jobs and new industry." [3]

Although $500 million may sound like a lot of money, in the business of orbital spaceflight it is, in reality, very little, due to the formidable challenges faced by a commercial launch operator that must build a vehicle capable of orbital insertion, rendezvous, and docking, each of which are several orders of magnitude more complex than the requirements of suborbital flight. However, for the few commercial space entities passionate about orbital flight, the COTS message was a godsend, and when, in March 2006, NASA opened the COTS program, 20 organizations submitted proposals, of which NASA selected 5 for further evaluation. On August 18th, 2006, NASA's ESMD announced SpaceX and Rocketplane Kistler as the two winners of Phase I of the COTS program, the former receiving $278 million and the latter being awarded $207 million.

WHO CAN AFFORD ORBITAL FLIGHT?

Based on studies conducted by the Futron Corporation [1] the first orbital spaceflight participants will most likely be male, aged early to mid-50s, be worth in excess of $200 million, and originate from the U.S.A, Europe, or the Asia/Pacific region [1]. Of the world's estimated 8 million millionaires, between 10,000 and 100,000 have the financial resources to consider an orbital space experience that is predicted to cost between $5 million and $8 million. This demographic subgroup understand risk but their time is valuable, a constraint that must be addressed when developing an orbital experience and the necessary training. For example, the first orbital spaceflight participants, Dennis Tito, Mark Shuttleworth, Anousheh Ansari, Greg Olsen, and Charles Simonyi, each spent several months training for their orbital flights, a situation that will clearly not be marketable if significant numbers of paying passengers are to fly into orbit [4]. A more realistic timeframe for orbital training is between five and six weeks, a duration that has been adopted when designing the training schedule presented in Chapter 6.

THE ORBITAL EXPERIENCE

The six preflight steps

Step #1

Those hoping to follow in the orbital footsteps of Dennis Tito and his fellow orbital spaceflight participants will first need to decide which company, agency, or space tour operator to fly with. Initially, this will be a relatively easy task, as there will only be two or three companies capable of offering an orbital flight experience, one of which is Space Adventures who arranged the flights of the aforementioned space passengers and the other is likely to be Bigelow Aerospace. Another possibility exists whereby a potential orbital spaceflight participant makes a suborbital trip with a company that is using suborbital flight as a stepping stone to orbital flight, such as SpaceDev. A part of your decision in choosing a company will be to evaluate the quality of training provided, the vehicle's mission architecture, landing/takeoff modes, orbital destination, and whether there is the option of a rendezvous with an orbital station.

Step #2

Having decided on an operator your next decision will be to choose which spaceport you wish to depart from, a decision driven by factors such as convenience, communications, specially developed space tourism facilities for the spaceflight participants' family and friends, and perhaps most importantly, the quality of the training facilities.

Step #3

Now that you have chosen the operator and departure location you will have the option of customizing your orbital experience. One of the options you will have will be to choose the type of orbital flight, such as a choice of orbital parameters, a rendezvous with an orbital habitat, or perhaps a stay on the ISS. Other personalization factors may include over-flight targets, for which you will be required to submit a list of "must see" features of Earth during the mission. No doubt, given the affluent clientele that will fly during the first few years of orbital operations, you will be provided with the option of being able to perform business-related activity during your stay on-orbit.

Step #4

Once you have paid your 10%, $500,000 deposit, you will need to submit to an extensive medical examination, which will be by several orders of magnitude more demanding than the medical criteria for suborbital flight. Because of the stricter medical requirements it is probable that some of those wanting to fly will be disqualified due to certain medical conditions that will be considered select-out criteria, the details of which are described in Chapter 5. The reasons for the more rigorous medical criteria are due to the greater acceleration and deceleration profiles experienced during typical orbital mission regimes. Because of the more demanding acceleration regime and the demands imposed on the cardiovascular system in particular, the medical examination will almost certainly require candidates to undertake a series of centrifuge runs in order to determine if any problems exist prior to launch.

Step #5

Details of the training schedule that you will be required to follow are described in Chapter 6. Although spaceflight participants will not be required to undertake the six months of training expected of their predecessors, the training will nevertheless be challenging and extensive in scope. A primary reason for this is the requirement that each spaceflight participant be proficient in emergency egress procedures, a situation that demands interdependency between each crewmember. Needless to say, training in safety and emergencies will form a major component of training, as will acquaintance training with the problems of microgravity. Unlike most suborbital flights, the launch and landing phases of an orbital trip to space will require each spaceflight participant to wear a pressure suit and helmet. Presently, such a suit costs $1 million, which means it is unlikely that you will be able to keep them as mementos without some financial consideration!

Step #6

Four to five days prior to launch, your families and friends will arrive at the spaceport to say goodbye prior to the quarantine period that will commence 96 hours before launch. During the quarantine phase, family and friends will be assigned a tour guide who will introduce them to the various space-themed educational and entertainment experiences available at the spaceport, such as the training simulators and the opportunity to experience microgravity onboard a zero-G aircraft. No doubt, the operator will hope that the experience offered will help sell the orbital experience and lead to subsequent flights by your family and friends!

THE FLIGHT

Launch phase

The flight phases will obviously vary depending on the operator's choice of mission architecture. One of the biggest differences in architecture will be whether the launch is vertical (such as atop a rocket stack as in Soyuz), horizontal, or air-launched (as envisaged by t/Space, which utilizes an aircraft during the first stage of flight). Another difference will be the size of the spacecraft and the number of crew and passengers. Bigelow Aerospace, for example, plans to carry up to five passengers in a capsule that will be launched into orbit using an Atlas V 401, whereas t/Space's CVX will be able to hold six passengers.

Orbital phase

This phase will also be architecture-dependent. In the case of Bigelow Aerospace, for example, the orbital phase will include a capsule phase followed by a habitat rendezvous and docking phase in which the Falcon 9 capsule will dock with Bigelow's BA330 module that will serve as an on-orbit vacation home. Other operators will require their passengers to remain in the capsule during their stay on-orbit. Regardless of which orbital mission architecture is utilized there will be certain aspects common to all. For example, routine activities such as food and drink preparation, sleeping arrangements, use of the toilet and shower facilities, laundry arrangements, and how to operate the communication links will each be a component of preflight training.

A major focus during time on-orbit will be exercise and entertainment, each of which will be architecture-dependent. For example, passengers onboard Bigelow Aerospace's BA330 module will have plenty of space to perform aerobatic and tumbling activities whereas passengers confined to a capsule will be more restricted. Since many of the early passengers will be businessmen it is likely they may wish to engage in business-related activities such as advertising or research, a topic addressed in Chapter 7. As the orbital flight industry evolves and matures, the opportunity for passengers to engage in EVA activities will be inevitable, but, based on the experience of professional astronauts it is likely that the most popular pastime on-orbit will be simply gazing out of the windows.

Leaving orbit

Perhaps the greatest physiological test faced by the inexperienced space traveler will be the challenge of leaving orbit due to the potentially distressing and deleterious effects of G forces imposed on the body during the descent. To mitigate against the potential cardiopulmonary injuries that re-entry may inflict on spaceflight participants it will be necessary for operators to provide effective training and to ensure that medical select-in and select-out criteria are adhered to rigorously.

After the flight

Shortly after being reunited with their families, spaceflight participants (now, newly minted civilian astronauts) will undergo a thorough postflight medical that will include blood tests, cardiopulmonary evaluation, and orthostatic evaluation by means of administration of the Stand Test. Based on the experiences of professional astronauts it is highly probable that you will experience backache and a certain level of discombobulation for several days following return to Earth due to the time it takes for the various physiological systems to re-adapt to 1 G.

The issuance of astronaut "wings" is a subject that has yet to be determined. It is possible that a new category of "spaceflight participant" wings will be awarded by the certifying authority of the FAA or it may be that the wings will be awarded by the operator itself. Before you depart the spaceport you will be given a multi-media disk containing a recording of your mission from both outside and inside the spacecraft. No doubt several of these multi-media sets will be provided to your family members and friends as they will serve as a very effective marketing tool!

ORBITAL FLIGHT RISKS

"Would I have flown if I had known there was a four percent chance of death? No, I don't think I would have flown. If I was told that I had a one percent chance of dying on the next space shuttle mission, I think I would take my chances."

Space Shuttle astronaut, Rick Hauck

Orbital flight risk assessment

One question that has evolved with the space tourism industry is how safe the vehicles will be for sending passengers into orbit. Although the public have remained steadfast in its support for the government-sponsored manned space program despite recent fatalities onboard the Space Shuttles Challenger and Columbia, such support may not extend to the private space industry in the event of an accident in which lives are lost. The problem for an orbital spaceflight participant is to determine what level of risk is acceptable. The 1% risk quoted by Rick Hauck may seem a little high when compared with the risk of dying in a commercial airline accident, but that risk level is in fact about the same that an average American has over his/her life of dying in a traffic accident. Of course, for many potential orbital spaceflight participants, the risk of flying into orbit is a part of the attraction, but for others it will be important to be able to make an informed assessment of that risk, which in turn may influence which operator they choose for their flight into orbit. Ultimately, although each spaceflight participant may determine their own individual level of risk, the determination will in reality be decided by the crews of future privately developed spacecraft, since it is only the crew that can assess whether all reasonable measures have been implemented to minimize risk.

Since many of the general suborbital considerations regarding which company to choose applies equally to orbital companies, there is no need to repeat them. However, given the much higher risk involved in orbital flight there are some pertinent questions the prospective orbital passenger may want to consider. For example, given the added complexities and significantly increased risks associated with orbital flight it is important to know what kinds of abort options are available during launch and re-entry and what the chances of survival are associated with each.

Abort procedures

The vehicle that will take you into orbit will feature a variety of abort modes of varying complexity depending on the severity of the situations that may be encountered during the various phases of flight. Some of the abort procedures will require only minor deviations from a nominal mission profile, whereas others may demand unusual, and in some cases, dangerous efforts by the crew and passengers. When evaluating the abort means of the vehicle it is important to understand the factors that may cause a failure of onboard systems and how these failures may degrade the vehicle's performance, since it is these parameters that may ultimately threaten the lives of the crew and passengers. For example, a failure of the vehicle's main engines will probably affect the vehicle's ability to reach orbit, but such a problem will probably not impact crew safety, whereas a cabin leak or a failure of the life support system would represent a major problem from which there may be few options for a successful outcome.

When making your evaluation you should investigate the vehicle's capabilities in being able to recover from a Pad Abort, an Intact Abort, and a Contingency Abort. A Pad Abort is an emergency situation in which circumstances prevent the vehicle from launching, whereas an Intact Abort occurs after launch but still permits the vehicle to return to the landing site for a safe recovery. Pad and Intact abort types have a high degree of automation that requires little crew effort to implement and often have a high probability of success, whereas the Contingency Abort is usually difficult to anticipate due to the circumstances of the vehicle's failure condition. Contingency Aborts, which require much more crew input, are the least desirable abort, and are generally less likely to succeed due to the unpredictable nature of the danger involved. For example, if your vehicle were to lose more than one main engine then it is unlikely it would be able to return to its landing site, and if it were to lose all engines then there may be few options open other than ditching or initiating an emergency egress.

Emergency Egress System

Another question you should ask is if the launch vehicle has an emergency egress system such as a launch escape system (LES), ejection seat, or a drag extraction system. An LES is a rocket mounted on the top of the crew capsule that is used to separate and launch the crew module away from the rocket in the event of an emergency, such as an event in which there is an imminent threat of a launch pad explosion, or during the initial ascent phase. In such an event the resulting ignition of the LES's escape rockets will subject the crew to an acceleration loading that may exceed 15 G for several seconds.

Given the violent acceleration loading, you may be wondering why rocket designers don't choose a more passenger-friendly emergency egress system, such as the ejection seat. Since the ejection seat generally subjects crewmembers to slightly less acceleration than a conventional emergency egress system and since it is also a much lighter system than a conventional rocket-based LES it would seem that this would be the egress system of choice for rocket designers. In addition to the advantages already mentioned, the ejection seat is also available as a means of emergency egress during both the ascent to orbit and the return to Earth and it is also a system that has been flight-tested in other rocket systems, such as the Russian Vostok, the American Gemini, and the test flights of the Space Shuttle Columbia. However, despite these seemingly obvious advantages, the use of ejection seats is not practical in vehicles that will carry the large crews that space tourism companies will be sending into orbit, since a separate seat and exit hatch must be provided for each crewmember.

The third, simplest, and least robust type of emergency egress system is the Drag Extraction System, which has been used on several experimental aircraft and which is currently being used as a means of emergency egress from the Space Shuttle. Such a system utilizes the air flowing past the vehicle to move the crewmember out of the capsule and away from the spacecraft using a special guide rail that explosively ejects from the vehicle. Before being able to use such a system the pilot would need to place the vehicle in a glide and bring the speed down to below Mach 1. Crewmembers would then have to unstow the Drag Extraction System, which looks like a telescopic pole that is attached to one of the vehicle's bulkheads. On the commander's order, a crewmember would then open one of the hatches and deploy the system. Each crewmember would then attach a lanyard to a hook on the pole and initiate the explosive eject system that would propel them clear of the vehicle. On reaching a safe altitude crewmembers would then deploy their parachutes. In reality, the chances of survival when using such a system are slim, but since the Drag Extraction System is the lightest and most uncomplicated system available there is a good chance that your vehicle will be fitted with some variant of it. If this is the case then you will be flying in a vehicle that has an escape system that cannot provide an egress capability during all phases of the launch and ascent profile and limited escape capability during the return to Earth.

COMPANY PROFILES

BIGELOW AEROSPACE

www.bigelowaerospace.com

"We need to encourage creativity, imagination and innovation in order to bring the benefits of space development to fruition, not just for the privileged few, but for all of humanity."

Robert. T. Bigelow, CEO, Bigelow Aerospace

Company profile

Robert T. Bigelow is a native of Las Vegas, Nevada, who made his fortune as a general contractor and founder of Budget Suites of America. In 1999 he founded Bigelow Aerospace, a general contracting, investment, and development company focused on realizing economic breakthroughs in the costs associated with the design, development, and construction of manned space habitats and launch facilities. Contrary to media reports Bigelow says he is not pursuing "space hotels", although he is interested in leasing his habitats, some of which may ultimately serve as a space hotel. To date he has spent more than $75 million of his own money and is prepared to invest up to $500 million by 2015 with the aim of realizing full-scale deployment of a manned space habitat. To achieve this Bigelow has obtained exclusive licenses on inflatable technologies and docking systems as well as multiple Space Act agreements with NASA.

"We consider ourselves wholesalers of destinations that we build and we don't consider ourselves as space hotel folks."

Robert T. Bigelow

The launch of BA's Genesis I Pathfinder (Figure 4.1) module in July 2006 and its successor, Genesis II in July 2007, represented a new chapter in the development and

Figure 4.1. Bigelow Aerospace's Genesis I Pathfinder inflatable module has been orbiting Earth since July 12, 2006. The launch put the company as many as five years ahead of a schedule that may result in fare-paying passengers embarking on orbital travel as early as 2010. Image courtesy: Bigelow Aerospace. Source: *www.thespacereview.com/article/710/1*

business of space, and signaled to the world that the orbital space race is no longer the exclusive domain of government aerospace industry.

Genesis I and II, the first privately funded space pathfinder modules, were sent into space during flawless launches using a Dnepr rocket from the ISC Kosmotras space and missile complex in Russia. The Genesis launches, combined with the developments at Bigelow's north Las Vegas center of operations, has put Bigelow as much as five years ahead of schedule, a situation that may result in a crew being launched as early as 2010. How did Bigelow do it?

Enabling technologies

Inflatable habitat technology

The core of Bigelow's operations is the use of inflatable habitat technology which forms the basis of each module. The initial subscale module, Genesis I, is the result of pioneering work by Bigelow, NASA, and various subcontractors in the development of lightweight but extremely strong and long-lived inflatable modules made of proprietary advanced aerospace materials. The inflatable module technology Bigelow Aerospace is using was originally a concept proposed as a crew quarters for the ISS. Known as the TransHab Project, the habitat technology was tested by Johnson Space Center (JSC) [6], but ultimately the project was canceled (in 2000) by NASA, although testing continues at JSC. The yet-to-be-flown manned module is a unique hybrid structure that combines the mass efficiency of an inflatable structure with the advantages of a load-bearing hard structure. It includes a bladder, a restraint layer, and micrometeoroid/orbital debris (MMOD) shell layers.

The module consists of almost two dozen layers that provide insulation against space temperatures that can range from $+121°C$ in the sun to $-128°C$ in the shade and protection against MMOD by the use of successive layers of Nextel (a material used as insulation under the hoods of cars) and several inches thick open cell foam. To test the shield, the University of Dayton Research Institute and the University of Denver Research Institute conducted more than 50 ballistic tests that fired particles of 0.6 cm to 1.25 cm toward the shield at velocities of between 3 km and 6.9 km per second. Other layers are composed of super-strong woven Kevlar that holds the module's shape and Combitherm, a material used for the bladder (of which there are three) construction [2].

The module's interior will be inflated to 10 psi (pounds per square inch), compared with 14.7 psi for the ISS and 12 psi for Skylab. To assess the habitat's structural integrity, several test articles have been manufactured and evaluated by JSC engineers. For example, one such evaluation involved testing to failure a 2.2-meter diameter restraint layer by increasing pressure to 197 psi (Figure 4.2), a test that demonstrated the ability to build a 10.5-meter diameter inflatable with a factor of safety (FOS) of 4.0 at 10.2 psi when using FAA Airship Requirements [7].

Another test investigated the structural integrity of a restraint layer when fitted with a window, an important feature for orbital spaceflight participants! Once again the restraint layer was inflated to an ultimate pressure of 197 psi before failing,

1.   2.   3.

4.

Figure 4.2. An 88-inch demonstration restraint layer being prepared for hydrostatic testing (photos 1, 2, and 3) at Johnson Space Center in November 2005. The restraint layer failed (photo 4) at 197 psi, demonstrating a factor of safety exceeding 4.0. Image courtesy: Bigelow Aerospace. Source: *http://ntrs.nasa.gov/archive/nasa/casi.ntrs.nasa.gov/20060022083_2006014634.pdf*

demonstrating that structural penetrations can be incorporated into the restraint layer without any reduction in strength.

Life support systems are being provided in part by established companies such as EADS Astrium and Boeing. For example, EADS Astrium is supplying a thermal/humidity control system and a carbon dioxide removal system that uses chemically coated beads instead of the molecular sieve technology on the ISS. A water handling system is being provided by Boeing. Several other subcontractors working for Bigelow are small to medium-sized aerospace companies that, by being involved with Bigelow, are gaining valuable aerospace experience and establishing a trend that will no doubt prove increasingly important as independent space travel becomes more common.

Bigelow Aerospace has already announced its launch manifest that will involve placing five modules into orbit, the first of which is the aforementioned Genesis I, a 4.4 m long spacecraft with 11.5 m^3 volume, eight solar arrays, communications antenna, a window, and six internal cameras. Genesis I currently orbits the Earth at 27,084 km/h (16,928 mph) and has an anticipated lifespan of between 3 and 13 years. It was joined in July 2007 by Genesis II, a spacecraft that features a multi-tank inflation system and augmented core attitude control and stabilization systems. Another upgrade was the addition of extra layers to the outer shield of Genesis II that better protect it against micrometeoroid collision damage. Joining Genesis I and II in the fall of 2008 will be Galaxy, a spacecraft with a 45% greater usable volume than its predecessors. Galaxy will serve as a test bed to bridge evolutionary development between Genesis I and II and the vehicles to follow. Galaxy will also be Bigelow's first human-habitable vehicle, providing critical risk reduction, and enabling first flight experience for technologies that will eventually be flown on Sundancer. These technologies will include advanced onboard avionics, an upgraded attitude determination and control system, a more robust air barrier that will be more damage-resistant, and higher performance battery technologies. A little more than a year following the launch of Galaxy, Sundancer will be launched onboard a SpaceX Falcon 9 booster to join the orbiting fleet of Bigelow spacecraft. With 175 m^3 of usable volume, Sundancer will be the largest Bigelow Aerospace spacecraft yet, and will feature life support systems proven onboard Galaxy and a robust propulsion system that will enable Sundancer to be launched into a high long-life orbit, before being maneuvered to lower altitudes where it will be able to be reached by manned spacecraft. Sundancer will also be launched with a Soyuz-type docking system and a new NASA-developed advanced lightweight Low Impact Docking System (LIDS). Finally, in 2012, BA will launch the BA330, the pride of its fleet of spacecraft, which will have a habitable volume of 330 m^3, sufficient to house up to six passengers and crew.

Mission architecture

Unlike SpaceX and Rocketplane Kistler, Bigelow Aerospace does not plan on building a man-rated launcher to fly spaceflight participants to its habitats. Instead, Bigelow will contract for flight services with various transportation providers and agree to buy a certain number of flights per quarter or per year. One of the transportation providers may be a winner of America's Space Prize, the orbital Ansari X-Prize equivalent, which challenges entities within the United States to design a reusable spacecraft, without government funding, capable of carrying passengers into orbit. The rules stipulate that the spacecraft must be capable of ferrying a crew and passenger complement of no fewer than five to an altitude of 400 kilometers and complete two orbits of the Earth. This feat must then be repeated within 60 days. The spacecraft must also demonstrate a docking capability with Bigelow's modules and be able to remain docked for up to six months. The competition deadline is January 10, 2010, with a cash prize of $50 million.

Figure 4.3. The Atlas V 401 configuration is the simplest, most robust, and most reliable version of the Atlas V and will probably be the launch vehicle that will ferry Bigelow Aerospace's first orbital passengers (see also color section). Image courtesy: United Launch Alliance. Source: *www. lockheed martin.com/data/assets/12462.pdf* or *www.aero.org/publications/crosslink/winter2004/07.html*

However, even Bigelow has acknowledged that the odds of a company claiming his prize are slim, which is why he is assessing the potential of the leading orbital companies to ferry his passengers to their habitats in space. It is possible that the first Bigelow Aerospace passengers will travel to LEO aboard a man-rated two-stage Atlas V 401 configuration, the simplest, most robust, and most reliable version of the Atlas V class (Figure 4.3, see also color section). The Atlas V, an intermediate-sized-class vehicle, is an upgraded Atlas launcher which was designed by Lockheed Martin as a part of its Evolved Expendable Launch Vehicle (EELV) program. The Atlas V variant first flew on August 21, 2002, when an Atlas V 402 successfully launched the Eutelsat Hot Bird 6 spacecraft from Cape Canaveral Air Force Station (CCAFS). The Atlas V is based on a common first-stage design known as the Common Core Booster™ and uses the NPO Energomash RD-180 engine. Since the Atlas V may be launched from either Launch Complex 41 at CCAFS or Vandenberg Air Force Base (VAFB) it is possible that Bigelow passengers may be the first humans to be launched from a site originally intended for launching military space shuttles.

Another potential launch provider is Rocketplane Kistler, with whom Bigelow Aerospace has signed a letter of intent regarding transportation to Bigelow's orbital habitats. The letter of intent states that once Rocketplane's K-1 vehicle is ready to carry passengers, and once Bigelow's habitats are in orbit, the two companies will do business to ferry passengers. However, given the financial and technological problems faced by companies such as Rocketplane Kistler and any other private space operator hoping to design a successful orbital vehicle and the potential for delays, Bigelow is keeping all his options open. On September 2006 Bigelow Aerospace entered into an agreement with Lockheed Martin to pursue the potential of launching passengers on human-qualified Atlas V rockets. The choice of the Atlas V is in keeping with Bigelow's vision, as he explained during the *Space 2006* conference.

> "We're a customer for whoever can produce an economical, reliable, safe transportation system that's user friendly. It's the other half of the coin. You have to have some place to go, but what good is an exotic island if there are no boats to get you there? One hand holds the other. We hope that over the next half-dozen years that as we go forward that, if we are able to make improvements and evolve towards full scale, other people will be doing something similar in this country."

At the Space Frontier Foundation's *NewSpace 2006* conference in Bigelow's home town of Las Vegas, he outlined his transportation plans once his full-scale habitat becomes operational.

> "In our third year of operation we estimate that we will need 20 launches. We need 16 launches for people and four launches for cargo. We've talked a lot about this over the last year and visited a lot with one company in particular and a second company secondarily and we would be most happy if a crew return vehicle and a hab vehicle for transportation purpose could carry eight people. The artifact of eight people is really a function of cost. It helps to reduce the seat cost. If you can still have an affordable seat cost and fly five people, that's fine. If you have a five-meter diameter on your crew return vehicle, you can accommodate eight folks in that architecture, and that can fly on a system like the Atlas V 401/402 series, or it could fly perhaps on the Falcon 9 if Elon [Elon Musk, founder of SpaceX] is of a mind to do that."

At 20 flights per year it is possible that Lockheed Martin will be able to offer flights as low as $35 million to $50 million, which would mean the costs per passenger would be as low as $5 million. Although Bigelow plans to fly many passengers per year he anticipates the major source of Bigelow Aerospace revenue will come from building space stations for research, manufacturing, and providing low-cost space programs to countries that do not have fully fledged space programs. For example, Bigelow is pursuing markets for a variety of users including biotech, pharmaceutical, university research, entertainment applications, and government and military users. At the *National Space Symposium* in April 2007, Bigelow announced prices for stays onboard his modules: for example, "Sovereign customers" (nations who want to send their astronauts into space) will be charged $14.95 million (in 2012 dollars) for a four-week stay in the BA330, the largest of the modules that will be launched, a ticket price that will include transportation to and from the habitat. Private companies that would like to lease a module for industrial research and/or processing will be charged $88 million per year for the BA330 module and $4.5 million a month for the half-size module. Reservations can be made with a 10% deposit that will be fully refundable until the launch of the subscale Sundancer module, initially scheduled for launch in 2010.

SPACE EXPORATION TECHNOLOGIES

www.spacex.com

"I think it's really incumbent upon us to extend life beyond Earth. Basically, to help make that happen is why I started SpaceX."

Elon Musk, CEO, Space Exploration Technologies

Company profile

Located in El Segundo, Southern California, near Los Angeles airport, Space Exploration Technologies (SpaceX) aims to offer light, medium, and heavy-lift capabilities to LEO using a suite of launch vehicles. SpaceX's founder, internet mogul turned spacecraft builder, Elon Musk, created the company in May 2002. Since then the company has already sold two vehicles, one to the U.S. military and the other to a foreign government. A winner of Phase I of NASA COTS competition, Musk's company received an award of $278 million for three flight demonstrations of its Falcon 9 rocket carrying the Dragon spaceship (Figure 4.4). These flights are scheduled to occur in late 2008 and 2009. The final flight will involve a transfer of cargo to the ISS and return cargo from the ISS back to Earth. The agreement between NASA and SpaceX includes an option for three demonstration flights of SpaceX's

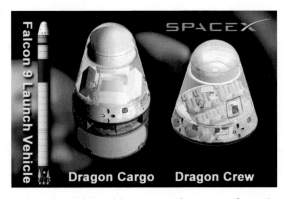

Figure 4.4. Dragon 9 launch vehicle with cargo and crew configurations. Image courtesy: SpaceX. Source: *www.spacex.com/photo_gallery.php*

seven-person manned version of Dragon, with the aim of using Dragon as a ferry to the ISS. In order to achieve rapid cargo to crew capability, the cargo Dragon and crew Dragon are almost identical, with the exception of the crew escape system, the onboard controls, and the life support system. If SpaceX is successful with its launch manifest it may win ISS resupply business worth between $300 million and $500 million per year after the retirement of the Space Shuttle in 2010. In addition to potentially servicing NASA's needs, the Dragon may also be of service to Bigelow Aerospace with whom SpaceX has an ongoing dialog to ensure that Dragon meets the human transportation needs of Bigelow's planned space station as efficiently as possible.

Enabling technologies

SpaceX's flagship launch vehicle is the Falcon 9, a two-stage, liquid oxygen and rocket-grade kerosene (RP-1)–powered rocket that measures 54.3 m (178 ft) in length and 3.6 m (12 ft) in diameter (Figure 4.5). The LEO mass with a 5 m fairing is an impressive 28,000 kg (716 klb) thanks to the vehicle's nine Merlin (hence the name "Falcon 9") engines being capable of generating a combined 4.4 MN (1 million lb) thrust. The Falcon 9 tank walls and domes are made from aluminum 2198 and are welded using friction stir welding, the most reliable welding technique. Equally robust is the interstage that connects the upper and lower stages, which is constructed of a carbon fiber–aluminum core composite structure. The separation system utilizes pneumatic pushers with pyrotechnic release bolts that will, following separation, detach the upper stage, which will be powered by one Merlin engine.

Much of the technology designed by SpaceX is driven by the findings of an Aerospace Corporation study that stated 91% of launch failures were due to engine problems, stage separation problems, or avionics. To minimize these problems, while still keeping costs low, SpaceX simply decided to design many of its own components, including its own breed of rocket engine. The result is the first-stage turbopump-fed Merlin engine that features several design elements common to other space-proven engines. For example, the pintle-style injector was initially used for the lunar landing engine during the Apollo Moon program. A single-shaft, dual-impeller turbopump operating on a gas generator cycle feeds propellant and also provides high-pressure kerosene for the hydraulic actuators. The turbopump also provides roll control by actuating the turbine exhaust nozzle. The vacuum-specific impulse of 304 s that the Merlin engine generates is more than impressive since it exceeds the performance of not only the Boeing Delta II main engine and the Lockheed Atlas II main engines, but also the fabled Saturn V F-1.

Figure 4.5. SpaceX's Falcon 9 rocket. Image courtesy SpaceX. Source: *www.spacex.com/falcon9.php*

The reliability of the Falcon 9 is reflected in its engine-out capability, which means that the vehicle can suffer the loss of one engine at any point in flight and still be able to complete the mission. Another reliability feature implemented into the Falcon 9's systems architecture is the hold-before-release system that results in the vehicle being held down after first-stage start and not being released until all propulsion and vehicle systems are operating nominally.

Dragon

The Dragon spacecraft comprises a nosecone, a pressurized section, and a service section in addition to a crew escape system. The capsule's 7-passenger up-crew capability is equal to its down-crew capability which permits the Dragon to function as an ISS shuttle and escape/lifeboat system. It carries 1,200 kg of propellant during a typical mission profile, an amount that is more than adequate for attitude control, ISS rendezvous, and orbital maneuvering. To mitigate against high Gs during re-entry and to facilitate landing precision, SpaceX's mission architecture is designed with a lifting re-entry. During re-entry, passengers are protected against thermal stress by an ablative, high-performance heat shield and sidewall thermal protection. During the latter part of 2008 SpaceX will conduct the first of three test flights of the Falcon 9, which will evaluate telemetry, communications, orbital maneuvering, thermal control, and re-entry capabilities. Two additional test flights are scheduled for 2009, the first of which will assess ISS rendezvous simulations and test the Dragon's approach, rendezvous, and breakaway operations. Finally, in the latter part of 2009 a full-cargo mission profile will be attempted, a flight that will include a docking with the ISS. SpaceX has already constructed a full-size prototype of the Dragon and tested its life support system.

Mission architecture

In launch configuration the Dragon capsule will sit atop the Falcon 9 vehicle, which will carry the spacecraft to a parking orbit of 185 km by 300 km. Although the Dragon will have an escape system fitted, SpaceX have yet to decide whether this will be a conical solid fuel escape tower or an integrated escape system that will use propellant intended for de-orbit. Once in orbit, the Dragon will flip its nose open to deploy the spacecraft's disk-shaped solar panels. At this stage the spacecraft's docking adaptor will be visible and the Dragon will commence its maneuver toward the ISS. Unlike the Space Shuttle, which maneuvers to a docking port under its own power, the Dragon will probably be maneuvered by the ISS's robotic arm, which will grapple the spacecraft and move to an appropriate docking port. Once attached to the ISS, Dragon may remain on-orbit for as long as six months before returning to Earth for a splashdown landing.

Spaceport

Falcon 9 launches of the Dragon capsule will take place from Cape Canaveral, thanks to an agreement between SpaceX and the U.S. Air Force. On April 26th, 2007,

SpaceX was granted a five-year lease by Air Force Space Command's 45th Space Wing to launch from Space Launch Complex (SLC) 40 located at Florida's Cape Canaveral Air Station, a site previously used for Titan 4 launches. As noted by Air Force General Kevin P. Chilton

> "The SpaceX license agreement is a good news story for the Air Force and nation. These types of agreements encourage entrepreneurial space achievement which can benefit both the Department of Defense and commercial space industries."

Although the terms of the license permits entities besides SpaceX to use the launch complex it is unlikely there will be any competition because, as Elon Musk noted, "Who else expects to build a heavy-lift launcher in that timeframe?" All six Falcon 9 launches that SpaceX has manifested will be flown from SLC 40, at least one of which will be a launch for Bigelow Aerospace.

ROCKETPLANE KISTLER AEROSPACE CORPORATION

Company profile

A feature common to each entrepreneurial aerospace company is the high level of expertise that makes up each team and in this regard RocketPlane Kistler (RpK) is no exception. Formed in 2001 to develop, build, and operate a vehicle capable of succeeding in the commercial suborbital and orbital market, RpK has assembled a formidable array of talent. Heading the company is Chief Executive Officer and Chairman, George French, a previous winner of the NASA AMES Research Astrobiology Team Group Achievement Award and the National Space Society's Entrepreneur of the Year Award. President, retired Colonel Randy Brinkley, served as NASA Program Manager for the ISS, as well as Mission Director of the Hubble Space Telescope Repair Mission.

Currently, RpK is in the process of developing the world's first commercial fully reusable orbital space vehicle, the two-stage K-1 spacecraft (Figure 4.6, see also color section). To achieve this RpK has, in addition to assembling its team of space experts, integrated its team with major contractors such as Honeywell, Irvin Aerospace, Oceaneering Space Systems, Lockheed Martin Space Systems, and Northrop Grumman Corporation. Another source of help is from NASA's COTS program for which RpK was selected as one of six finalists in March 2007.

Enabling technologies

To ensure reusability, the K-1 incorporates existing technologies that have been adapted from previously successful aerospace programs. For example, the K-1's engines, which form the heart of the Kistler K-1 and are used for both first and second stages, were originally developed and built for the Soviet Union's N-1 rocket designed for the Russian Moon program more than 30 years ago. The Soviet Union

Figure 4.6. Kistler Aerospace's K-1 vehicle (see also color section). Image courtesy: Kistler Aerospace. Source: *www.rocketplanekistler.com/photogallery/wallpaper/large/inflight.html*

abandoned the N-1 program in 1974 and placed about 100 N-1 engines (originally designated NK-33 and NK-43) in storage. More than 30 years later, Aerojet of Sacramento reached an agreement with the constructors of the NK-33 and NK-43 engines and imported 46 of them for use on the K-1. After being modified with U.S. ignition systems, control valves, and thrust vector control (TVC) systems to make them reusable, the engines were renamed, the NK-33 becoming the AJ26-58, and the NK-43 becoming the AJ26-59. According to Aerojet, the modified engines have a 99.5% reliability rating. The K-1 vehicle that will utilize the engines is a two-stage, fully reusable spacecraft, 36.9 m in height, 6.7 m in diameter, and weighing 382,300 kg. The first stage, which is also known as the Launch Assist Platform (LAP), is 18.3 m high and is fitted with three liquid oxygen (LOX)/kerosene engines (two AJ26-58s and one AJ26-59). The second stage, which measures 18.6 m in height, includes the extended module that is the actual Orbital Vehicle. The 4,540 kN (1,020,000 lbf) combined thrust of the LAP engines, each capable of being hydraulically gimbaled to ±6°, is similar to the thrust of SpaceX's Falcon 9, and will, when combined with the 1,760 kN thrust of the Orbital Vehicle engine, be able to place the payload module into the 220 km orbit necessary for ISS-servicing missions.

Reusability of the K-1 is assured by designing structural components with a large factor of safety. For example, the K-1 uses military-grade, radiation-hardened global-positioning systems and a triplex fault-tolerant avionics architecture. The Integrated Vehicle Health Management System (IVHMS), which is used to perform the preflight checkout, is similar to systems fitted onboard commercial aircraft such as the Boeing 777.

Initially, RpK plans to use the K-1 reusable spacecraft to deliver payloads to LEO, some of which may be destined for the ISS. One of the advantages of the K-1 vehicle configuration is its versatility thanks to its capability of being fitted with an expendable third stage that allows RpK to offer payload delivery to high-energy orbits. With the retirement of the Shuttle in 2010 and the projected increase in the ISS crew size to six, the demand for cargo resupply, cargo recovery, re-boost services and crew will increase. Thanks to the vehicle's versatility, the K-1 is able to fulfill each of these demands simply by reconfiguring payload modules. For example, for an ISS resupply mission the K-1's payload module would be replaced with the ISS Cargo Module that is capable of carrying experiment racks, equipment, and supplies. Alternatively, for a Bigelow mission, the payload module would be replaced with a pressurized crew module, designed to carry up to four passengers and crew.

Mission architecture

The K-1 will reach stage separation altitude of 43.2 km 121 seconds after launch, at which point the K-1 will be traveling at 1.22 km/s. At this altitude the first stage will separate from the Orbital Vehicle and 4.4 seconds after separation, the LAP center engine (the AJ26) will re-ignite to return the stage to the launch site on a controlled return trajectory. At an altitude of 3,030 m, six parachutes will deploy and shortly prior to landing, four airbags will inflate to cushion the landing so the first stage can be refurbished and prepared for its next launch.

The OV main engine (the former NK-43) will ignite 7.3 seconds after stage separation and continue to burn until the K-1 vehicle attains an altitude of 94.4 km at which point it will be traveling at 7.8 km/s. Following OV engine cutoff, the OV will coast to the target altitude before the OMS will engine-ignite to circularize the orbit. Approximately 60 minutes after launch, the K-1 will be in its intended orbit and approximately 30 minutes following orbital insertion the payload will be deployed.

Spaceport

Location. Kistler will operate its K-1 vehicle from two launch sites, one of which is Spaceport Woomera, located at Woomera, Australia, about 470 km north of

vative crew seats that have the ability to rotate 180° for entry or in the event of an abort. Other safety features include the CXV exterior, which will be covered with two layers of ablative Silicone Impregnated Refractory Ceramic Ablator (SIRCA) tiles.

Since the CXV will be air-launched, t/Space is also involved in the design of a Very Large Aircraft (VLA) which will carry the capsule and its booster underneath in a similar approach to the one adopted by Scaled Composites' WhiteKnight carrier aircraft and its suborbital vehicle SpaceShipOne. An alternative to the VLA will be to modify the landing gear of a B-747 aircraft to ensure sufficient ground clearance for the capsule and its booster.

Mission architecture

t/Space is unique among current private space companies in its use of what is termed a Trapeze-Lanyard Air Drop (t/LAD) launch method, a concept conceived in December 2004 by Dr. Marti Sarigul-Klijn [5], Program Manager for t/Space's air launch test program. The t/LAD air launch trajectory bears some similarities with Scaled Composites' WhiteKnight/SpaceShipOne mission architecture, but whereas SS1 drops from its carrier vehicle and then re-crosses in front of it, t/Space's CXV employs an aft-crossing trajectory (Figure 4.9, see also color section). During the first phase of the mission the CXV will be slung under a carrier aircraft such as a modified B-747 airliner or from a custom VLA. At the drop altitude the t/LAD mechanism holds onto the nose of the CXV after the booster is released so that it slowly rotates into a vertical orientation while a parachute, which is used to slow the rate of rotation, simultaneously deploys from the back of the booster. Since the drop altitude is between 7,500 m and 9,090 m, t/Space's architecture provides safe abort modes during the critical first seconds of flight as there is no requirement for the enormous energies demanded of a ground-launched escape system. Another abort advantage of an air-launch configuration is the simple fact that the capsule is already high enough for parachute deployment.

t/Space's air launch architecture simplifies the process of launching into orbit in a number of ways. First, the constraints imposed by weather conditions are reduced by flying to altitude and also by the fact that if weather impacts the planned launch area the vehicle can always be flown to an alternate launch point. Second, since ground-launch facilities are not used, there is no impact by ships entering ocean zones or concerns of rocket stages dropping. The capsule will be designed to land on water via parachute, a system that has already been tested using a full-size and full-weight Drop Test Article (DTA) of the CXV. In August 2005, t/Space released a DTA from a Sikorsky S-61 helicopter at an altitude of 2,850 m to test the triple-parachute system and the deceleration profile of the vehicle during a typical water entry. Based on the successful test it is anticipated that the CXV's passengers will experience 4.5 G on re-entry, a deceleration greater than the Space Shuttle, but less than the Gemini capsule.

On a budget that is normally barely sufficient to create a PowerPoint presentation at NASA, t/Space has built a full-scale mockup, demonstrated a unique method of air-launching its vehicle and successfully drop-tested a vehicle during a parachute deployment and capsule recovery test. With its technological innovations and its

Figure 4.9. The release of t/Space's test article from a Proteus carrier plane shown in a sequence of images taken at half-second intervals (see also color section). Image courtesy: Dr. N. Sarigul-Klijn/Kistler. Source: *www.space.com/php/multimedia/imagedisplay/img_display.php?pic= v_cxv_dropsequence_02.jpg&cap*

"faster, cheaper" development ethos, t/Space has proved itself as a credible company that may just convince NASA to part with the $400 million it needs to build the CXV and have it operational before the retirement of the Shuttle.

REFERENCES

[1] Beard, S.S.; and Starzyk, J. *Futron/Zogby Space Tourism Market Study: Orbital Space Travel and Destinations.* Futron Corporation, October 2002.
[2] Gossamer Spacecraft, in *Membrane and Inflatable Structures Technology for Space Applications.* American Institute of Aeronautics and Astronautics (Progress in Astronautics and Aeronautics, Volume 191, 2001).

[3] Griffin, M. American Astronautical Society, Speech Excerpt, November 2005.

[4] Leonard, D. *Space Tourism: Marketing to the Masses*. Space.com. Online at *www.space. com/adastra/050606_isdc_tourism.html*. June 6, 2005.

[5] Sarigul-Klijn, M.; Sarigul-Klijn, N.; and Noel, C. Air-Launching Earth-to-Orbit: Effects of Launch Conditions and Vehicle Parameters. *AIAA Journal of Spacecraft and Rockets*, **42**(3), 2005.

[6] *Transhab: NASA's Large-scale Inflatable Spacecraft*. American Institute for Space Aeronautics and Astronautics, 3/06/2000.

[7] United States Patent, Inflatable Vessel and Method, 6,547,189. April 2003.

5

Medical certification: Spaceflight participant medical standards and certification

THE SPACEFLIGHT ENVIRONMENT

The medical certification requirements for orbital flight are by several orders of magnitude more demanding and restrictive than those for suborbital flight since most, if not all physiological systems, are affected by microgravity, some more seriously than others. In addition, significant pathophysiological effects have been observed in astronauts, such as crewmembers being incapacitated for several days early in a mission due to repeated vomiting associated with space motion sickness. Other astronauts have suffered serious postflight orthostatic intolerance induced by cardiovascular deconditioning [2, 13, 21], and yet others have experienced significant bone loss.

Before describing the medical requirements for orbital flight it is important to be cognizant of the extent to which even short exposure to microgravity exerts its influence on the human body. After all, since you will be spending U.S.$5 million for a week in orbit, you should at the very least understand the consequences of the trip on your body and why you will be required to submit to what will be an extensive, prolonged, and fairly uncomfortable medical examination. To provide you with the rationale for the medical requirements and to describe briefly the effects that microgravity exposure will have on you, this section describes how the spaceflight environment affects the various physiological systems of the body and the pathophysiological effects that occur.

Physiological vs. pathophysiological effects

Differentiation between the physiological and pathophysiological effects of spaceflight is difficult, if not impossible, since transitions between the different states fluctuate and are ill-defined. Also, whether changes in body function and structures are called physiological or pathological depends mainly on the reference condition.

Thus, adaptive responses during spaceflight may be essential for survival in the microgravity environment and therefore may be called physiological. However, on returning to Earth, the same effects can lead to a breakdown in the body's integrity and become pathological.

The physiological adaptations that occur in the body and the specific effects on each physiological system are described in this section.

Cardiovascular system

One of the first cardiovascular changes you will experience following orbital insertion is a headward fluid shift of between 1.5 L and 2.0 L and an accompanying increased heart rate that may be almost double your resting heart rate. The major sources of the fluid shifted are the lower extremities. This fluid shift causes venous distension, facial edema, and the feeling of nasal stuffiness in addition to an increased orthostatic intolerance during the first week of spaceflight. This is generally followed by post-flight syncope [6, 21], which means that you will feel light-headed on your return to Earth. Fortunately, although the cardiovascular parameters you will experience will change throughout your flight and may take several weeks to return to baseline, eventually all parameters will return to their set points. For example, on your return to Earth, these cardiovascular adaptations will cause you to notice your resting heart rate to be higher than normal and you may feel slightly discombobulated for the first two or three days, but within a week of your return you will feel no different than you did before the flight.

Bone and muscle metabolism

During your stay on orbit, the mechanical load that your bones must support will be reduced to almost zero, a situation compounded by the fact that many other bones that you use to aid in movement are used less intensively in microgravity. Gradually, even if you stimulate your bones by artificial mechanical loading, you will experience disuse osteoporosis and muscle atrophy. This occurs because your bones are relieved of mechanical loads, which causes the calcium, which gives your bones their strength, to be excreted via the bloodstream, resulting in a loss of density and bone mass. The bad news is that this process begins as soon as you enter orbit and continues indefinitely! The good news is that during a short-duration mission of between 7 and 14 days you will not decrease your fracture threshold on your return to Earth. However, if you embark on a long-duration mission, be advised that space medicine experts are still unsure how the mechanism of bone loss occurs in microgravity and, more importantly, they do not understand how to prevent bone loss from occurring. For example, there have been cases of astronauts (such as Thomas Reiter, who spent six months onboard the Russian space station Mir), who, despite implementing extensive countermeasures designed to negate the effects of bone loss, suffered more than 10% loss of the bone mass density in the bones of their lower legs. Ten years after the mission, some of these astronauts have yet to regain their bone density!

There is also the not inconsequential effect on your muscles, since the absence of mechanical load will have a similar atrophying effect on your muscle mass. As muscles adapt based on their loading histories, an absence of load, such as your body will experience in microgravity, will impose significant degradation rates in your muscle volume, protein synthesis, and lean body mass. This gradual loss of muscle function will occur regardless of how intensive your countermeasure regime is, or how many pharmacologic intervention strategies you implement. Generally, during a typical short-duration mission, you can expect to suffer a 15% reduction in the cross-sectional area of your slow-twitch muscle fibers and a 22% reduction in the cross-sectional area of your fast-twitch fibers, which will mean that you may feel a little weak when you perform exercise during the first week on return to Earth. The good news is that, after a couple of weeks your muscles should have recovered most of their strength.

Hematological and immunological parameters

Following your return to Earth you may experience shortness of breath for a few days since your body will have lost red blood cell mass (RBCM). The mechanism for this is directly related to your body's plasma volume (PV) and is one of the adaptations to microgravity that is fairly well understood. Due to the fluid distribution that occurs on orbital insertion, your body will perceive there is too much fluid and it will therefore excrete water, resulting in a concomitant reduction in plasma volume, since much of plasma is water. This decrement in PV then causes an increase in hemoglobin concentration, which you would think would be a benefit, but this is not the case in microgravity! The increase in hemoglobin concentration effects a decrease in erythropoietin (EPO) and, due to the destruction of recently formed RBCs, the RBCM ultimately decreases to a level that the body determines as acceptable for a microgravity environment. Fortunately, although PV decreases soon after exposure to microgravity and remains low throughout the mission, it generally returns to baseline between 7 and 14 days following return to Earth.

Other adverse effects you may experience will be those related to your immune system. The inflight psychological, environmental, and physiological stressors associated with your orbital stay will impose adverse effects on your immune systems and may lead to you becoming more susceptible to latent virus infections such as Epstein–Barr virus and an overall decreased recovery time following your return to Earth [21]. Examples of the most significant spaceflight factors that may influence your immune system include disrupted circadian rhythm and sleep disruptions, microbial contamination, and psychological stressors such as the isolation that may be experienced in missions extending to more than two weeks.

Endocrine changes

Less overtly noticeable are the changes that occur in the endocrine system. The redistribution of blood throughout the vasculature that occurs as a result of microgravity causes the endocrine system to respond by changing hormone output. Since the alteration in blood volume is interpreted as a relative volume expansion, this fluid redistribution necessitates a compensatory change in water balance with a net loss of fluid and electrolytes, thereby upsetting the hormonal balance in your body and resulting in weight loss. The extent of this fluid and electrolyte loss and hormone disruption is related to food consumption and changes in water balance, which occurs principally in the first or second day of flight. Fortunately, the endocrine system adapts quickly, and you regain much of your weight loss within the first 24 hours following return to Earth, and the remainder within a week postflight.

Neurovestibular system

Perhaps some of the more unpleasant adaptations that you will experience during your flight are the ones affecting the neurovestibular system, which is the system responsible for balance, posture, and spatial awareness. Unsurprisingly, in microgravity, where there is no "up" or "down", this system is forced to deal with unusual and conflicting signals that result in a sensory mismatch between inputs to the system. Due to the initial disorientation that occurs as a result of the neurovestibular system adapting to the new microgravity environment, more than 60% of first-time space travelers experience disturbances in postural control, which may include a variety of vestibular reflex phenomena such as rotation, vertigo, and movement illusions. These disturbances ultimately manifest themselves in typical motion sickness symptoms that range from headache and stomach awareness to nausea and vomiting. The symptoms normally begin shortly after entry into orbit and may be triggered by viewing an unusual scene, such as an inverted crewmate; although even minor head movements often provoke them. Unfortunately, despite the obvious operational hazards associated with space motion sickness and despite decades of research investigating the syndrome, no single drug or combination of drugs has been proven to protect astronauts. Fortunately, unlike some other physiological systems, the neurovestibular system usually adapts within 72 hours, although the rate of recovery, degree of adaptation, and specific symptoms vary widely among individuals. Until your symptoms resolve, you will be advised to avoid rapid head movements and your flight surgeon will recommend that you administer transdermal Scopolamine or oral Promethazine, the latter being the current recommended pharmacologic countermeasure of choice.

Psychiatric

> "All the necessary conditions to perpetrate a murder are met by locking two men in a cabin of 18 by 20 feet ... for two months."
>
> Cosmonaut, Valery Ryumin

Isolation is perhaps the most inescapable of all stressors in spaceflight, and one that has been graphically depicted in such Hollywood movies as *Solaris*, *Sunshine* and *Silent Running*. While no space mission has ended in murder, isolation can lead to

Table 5.1. Psychological stressors of short-duration spaceflight.

Psychological	Psychosocial	Human factors
Isolation and confinement	Coordination demands	Workload
Limited possibility for abort and rescue	Interpersonal tension between crew	Limited exchange of info/communication with external environment
High-risk conditions and potential for loss of life	Enforced interpersonal contact	Limited equipment, facilities, and supplies
Hostile external environment	Crew factors (i.e., gender, size, personality, etc.)	Mission danger and risk associated with equipment failure, malfunction, or damage
Alterations in sensory stimuli	Social conflict	Food limitations

numerous problems including sleep deprivation, depression, irritability, anxiety, impaired cognition, and even hostility. To the isolation, you can add noise, odors, crew personalities, atmosphere, diet, changes in circadian rhythm, and the fatigue and lassitude that eventually affect every space traveler to some degree. These latter stressors can be considered latent and are caused by the unavoidable realities of spaceflight, but crewmembers may also be faced with overt stressors, which arise from specific events such as critical failures of equipment [4, 8, 9, 11, 12, 14].

Although many of the problems may be manifested by astronauts deployed on long-duration missions, it should be emphasized that, with the exception of five orbital spaceflight participants (who, despite their status as tourists, still received six months of training), those astronauts who have spent time on orbit have, without exception, had extensive training in learning how to deal with the problems outlined in Table 5.1. Given that future orbital spaceflight participants will receive less than a third of the training that previous paying passengers received, the potential for psychological stressors during short-duration flights must be expected.

It will therefore be imperative that operators implement a process that will ensure that orbital spaceflight participants are psychologically fit to fly.

Radiation

During your stay on orbit you will be exposed to levels of radiation that far exceed terrestrial norms. Although radiation represents the primary hazard to orbital spaceflight participants and despite extensive data collection and research over the past three decades, little is known about the degree, consequences, and potential amelioration of possible damage to crewmembers. A given radiation exposure is dependent on the trajectory and duration of stay on orbit as well as the amount of shielding present. However, due to the complex spatiotemporal variations in space radiation, combined with the non-uniform distribution of spacecraft contents, it is very difficult to accurately quantify and measure exact radiation exposure values. However, because of the protective effect of the Earth's magnetic field and the low-altitude, low-inclination orbits that will probably be utilized by your operator, the problem of excessive radiation exposure should be minimized.

Toxicological hazards

Toxic substances in your vehicle may originate from a number of different sources, including leaks or spills, volatile metabolic waste products, thermodegradation of vehicle materials and products, and particulate pollutants. The route of absorption and exposure can be via inhalation, ingestion, or direct contact. Since these contaminants have the potential to impose significant stress on the health and wellbeing of crewmembers, several hours of the orbital training program are dedicated to instructing spaceflight participants in the appropriate emergency responses to a spillage of any hazardous material [5]. Some examples of the toxicological hazards you may be exposed to are provided in Table 5.2. The table lists guideline values that have been determined using information provided by the National Research Council (NRC) Committee on Toxicology and take into account the unique factors associated with spaceflight environment stress on human physiology and the health of crewmembers. Documentation of the values in this section can be found in a four-volume series of books entitled *Spacecraft Maximum Allowable Concentrations for Selected Airborne Contaminants* published by the National Academy Press, Washington, D.C. (*www.nap.edu/info/browse.htm*).

Table 5.2. Toxicological hazards of the spaceflight environment.

	POTENTIAL EXPOSURE DURATION			
	24 hour		7 days	
Chemical name	**ppm**	(mg/cm^3)	**ppm**	(mg/cm^3)
Ammonia Synonyms: NRC Vol. #1	**20**	(14)	**10**	(7)
	Organ Mucosa	*Effect* Irritation	*Organ* Mucosa	*Effect* Irritation
Benzene Synonyms: NRC Vol. #2	**3**	(10)	**0.5**	(1.5)
	Organ Blood	*Effect* Immunotoxicity	*Organ* Blood	*Effect* Immunotoxicity
Carbon monoxide Synonyms: NRC Vol. #1	**20**	(23)	**10**	(11)
	Organ CNS	*Effect* Depression Arrhythmia	*Organ* CNS Heart	*Effect* Hyperventilation Arrhythmia

Chemical name	POTENTIAL EXPOSURE DURATION			
	24 hour		**7 days**	
	ppm	(mg/cm^3)	**ppm**	(mg/cm^3)
Dichloroacetylene Synonyms: NRC Vol. #3	**0.04**	(0.16)	**0.03**	(0.12)
	Organ CNS Liver Kidney	*Effect* Depression Hepatotoxicity Nephrotoxicity	*Organ* CNS Liver Kidney	*Effect* Depression Hepatotoxicity Nephrotoxicity
Ethylbenzene Synonyms: NRC Vol. #1	**60**	(250)	**30**	(130)
	Organ Mucosa CNS	*Effect* Irritation Depression	*Organ* Mucosa Testes	*Effect* Irritation Necrosis
Ethylene glycol Synonyms: NRC Vol. #1	**25**	(60)	**5**	(13)
	Organ Mucosa CNS	*Effect* Irritation Depression	*Organ* Mucosa CNS Kidney	*Effect* Irritation Depression Nephrotoxicity
Freon 11 Synonyms: NRC Vol. #4	**140**	(790)	**140**	(790)
	Organ Heart	*Effect* Arrhythmia	*Organ* Heart	*Effect* Arrhythmia
Freon 22 Synonyms: NRC Vol. #4	**1,000**	(3,500)	**1,000**	(3,500)
	Organ CNS Heart	*Effect* Depression Arrhythmia	*Organ* CNS Heart	*Effect* Depression Arrhythmia
Glutaraldehyde Synonyms: NRC Vol. #4	**0.04**	(0.08)	**0.006**	(0.025)
	Organ Mucosa	*Effect* Irritation	*Organ* Respiratory system	*Effect* Lesions
Hydrazine Synonyms: diamine NRC Vol. #2	**0.3**	(0.4)	**0.04**	(0.05)
	Organ Liver	*Effect* Hepatotoxicity	*Organ* Liver	*Effect* Hepatotoxicity
Hydrogen cyanide Synonyms: NRC Vol. #1	**4**	(4.5)	**1**	(1.1)
	Organ CNS CNS CNS	*Effect* Depression Headache Nausea	*Organ* CNS CNS CNS	*Effect* Depression Headache Nausea

(*continued*)

Table 5.2 (*cont.*)

	POTENTIAL EXPOSURE DURATION			
	24 hour		**7 days**	
Chemical name	**ppm**	(mg/cm^3)	**ppm**	(mg/cm^3)
Indole Synonyms: NRC Vol. #2	**0.3**	(1.5)	**0.05**	(0.25)
	Organ CNS Blood	*Effect* Nausea Hematoxicity	*Organ* CNS Blood	*Effect* Nausea Hematoxicity

Methane
Synonyms:
  natural gas
  NRC Vol. #1
	5,300	(3,800)	**5,300**	(3,800)

	Organ	*Effect* Explosion	*Organ*	*Effect* Explosion

Methylene chloride
Synonyms:
  dichloromethane
  NRC Vol. #2
	35	(120)	**15**	(50)

	Organ CNS	*Effect* Depression	*Organ* CNS	*Effect* Depression

Octamethyl trisiloxane
Synonyms:
  MDM
  NRC Vol. #1
	200	(2,000)	**100**	(1,000)

	Organ Liver Kidney	*Effect* Death Hepatotoxicity Nephrotoxicity	*Organ* Liver Kidney	*Effect* Hepatotoxicity Nephrotoxicity

2-Propanol
Synonyms:
  isopropanol
  NRC Vol. #2
	100	(240)	**60**	(150)

	Organ CNS Mucosa Liver	*Effect* Depression Irritation Hepatotoxicity	*Organ* CNS Mucosa Liver	*Effect* Depression Irritation Hepatotoxicity

Toluene
Synonyms:
  methyl benzene
  NRC Vol. #2
	16	(60)	**16**	(60)

	Organ CNS	*Effect* Depression	*Organ* CNS Mucosa	*Effect* Depression Irritation
Trichloroethylene Synonyms: NRC Vol. #3	**20**	(74)	**10**	(37)
	Organ CNS	*Effect* Depression	*Organ* Kidney Liver	*Effect* Nephrotoxicity Hepatotoxicity

Vinyl chloride
Synonyms:
  chloroethene,
  chloroethylene
  NRC Vol. #. 1
	30	(77)	**1**	(2.6)

	Organ Liver CNS CNS	*Effect* Hepatotoxicity Headache Depression	*Organ* Testes Liver	*Effect* Necrosis Hepatotoxicity

CNS = Central Nervous System.

Medical standard	Spaceflight participant
Distant vision	20/40 or better separately, with or without correction
Audiology (must pass either test)	Audiometric speech discrimination test score of at least 70% in one ear. Pure tone audiometric test, unaided, with thresholds no worse than

Hz	500	1,000	2,000	4,000
Better ear	35 dB	30 dB	30 dB	40 dB
Worse ear	35 dB	50 dB	50 dB	60 dB

Medical standard	Spaceflight participant
Ear, nose, throat (ENT)	No ear disease or condition manifested by, or that may be reasonably expected to be manifested by, vertigo or a disturbance of speech or equilibrium
Pulse	Used to determine cardiovascular system status and responsiveness
Electrocardiogram	Not required
Blood pressure	No specific value stated. Recommended possible treatment if BP is on average greater than 155/95 mmHg
Psychiatric	No diagnosis of psychosis, bipolar disorder, or severe personality disorder
Substance dependence or substance abuse	A diagnosis or medical history of substance dependence or abuse will disqualify you unless there is established clinical evidence, satisfactory to the flight surgeon, of recovery, including sustained total abstinence from the substance(s) for 2 years preceding the examination. Substance includes alcohol and other drugs (i.e., PCP, sedatives and hypnotics, anxiolytics, marijuana, cocaine, opiates, amphetamines, hallucinogens, and other psychoactive drugs or chemicals)
Disqualifying conditions	An SFP will be disqualified if they have a history of (1) diabetes mellitus requiring hypoglycemic medication; (2) angina pectoris; (3) coronary heart disease that has been treated or, if untreated, that has been symptomatic or clinically significant; (4) myocardial infarction; (5) cardiac valve replacement; (6) permanent cardiac pacemaker; (7) heart transplant; (8) psychosis; (9) bipolar disorder; (10) personality disorder that is severe enough to have manifested itself by overt acts; (11) substance dependence; (12) substance abuse; (13) epilepsy; (14) disturbance of consciousness without satisfactory explanation of cause; or (15) transient loss of control of nervous system function without satisfactory explanation of cause

(*continued*)

Table 5.3 (*cont.*)

Medical standard	Spaceflight participant
Medical history	As outlined in SPMC
Physical examination	Includes rectal examination, pelvic examination, and proctosigmoidoscopy
Cardiopulmonary evaluation	Includes history and examination, pulmonary function tests, exercise stress tests, blood pressure, resting ECG, and echocardiogram
Musculoskeletal evaluation	Muscle mass and anthropometry
Radiographic evaluation	Chest films (PA and lateral), sinus films, mammography, and review of medical radiation exposure history
Laboratory examinations	Complete blood workup (clinical biochemistry, hematology, immunology, serology. and endocrinology) Urinalysis, including urine chemistry and renal stone profile Stool analysis of occult blood, ova, and parasites
Otorhinolaryngo-logic (ENT) evaluation	Includes history and examination, audiometry, and tympanometry
Ophthalmologic evaluation	As outlined. No further testing required
Dental examination	Includes panorex and full dental X-rays within prior 2 years
Neurological evaluation	Includes history and examination EEG at rest EEG with photic stimulation EEG during hyperventilation
Psychiatric and psychological evaluation	Includes psychiatric interview and psychological tests
Other tests	Drug screen Microbiologic, fungal, and viral tests Pregnancy test Screen for sexually transmitted disease Abdominal ultrasound

Adapted from *FAA Guide for Aviation Medical Examiners* [23].

Outcomes of your medical examination

Before discussing how best to prepare for what will probably be the most rigorous and extensive medical examination of your life it is important to understand the possible outcomes of the examination. This section describes the possible results of your medical examination and the steps you should take in the event that you are denied or deferred your medical certification.

Issuance, denial, and deferral

Issuance. Upon arrival at the medical facility you will be required to submit your report-of-medical-history form, after which the physical examination will be conducted by an FAA-certified flight surgeon with extensive experience in the field of aerospace medicine. Following the examination, providing you meet the standards, you will be issued a provisional spaceflight participant medical certificate.

Denial. If you do not meet the standards required for orbital flight your flight surgeon will forward your information to the operator's flight surgeon. The flight surgeon will then issue a denial letter stating that you are medically disqualified from orbital flight. On receipt of this letter you will have up to 30 days in which to appeal the decision. This will require you to present medical information corroborating that the disqualifying medical condition is safe, a process that will require you to be examined by an aerospace surgeon who specializes in the physiological system associated with the disqualifying condition. On submission of request to overturn the waiver, the operator's flight surgeon will assess your medical file and make a medical recommendation to the approval authority.

Deferral. There will be certain conditions which will be subject to waiver provided that you meet specific requirements. If this is the case, the examining physician will defer your application to the operator's flight surgeon, indicating the nature of the condition(s) requiring consideration for waiver. You will then be required to be examined by an aerospace surgeon who specializes in the deferral condition. Assuming you meet the waiver requirements, certification will be approved. If you do not meet the waiver requirements you will be issued a letter of denial and have 30 days in which to appeal the decision by submitting a request in writing to the flight surgeon. A more detailed account of the waiver process is outlined below and in the algorithm depicted in Figure 5.1.

Waiver process

This process ensures the consistent and proper management of those whose medical eligibility has been deferred. The process will normally start at the examining physician's clinic at the time of the discovery of the disqualifying condition. Local evaluations and consultations will be performed and the individual's condition will be documented in a Spaceflight Medical Summary (SMS), which will then be forwarded to the flight surgeon.

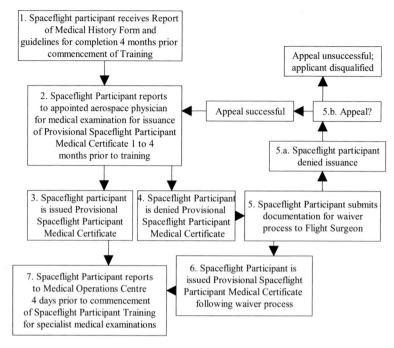

Figure 5.1. Algorithm of the waiver process.

It is expected that the majority of waiver requests will be routine waivers that will have clear policy established and require little review and endorsement. The SMS and supporting disposition letters will be reviewed by the flight surgeon, who will then make a recommendation to the approving authority. In certain cases, the flight surgeon and/or the approving authority may request a second opinion or additional consultation(s). This will usually consist of a case review which will require the individual's SMS and supporting documentation to be sent to another aerospace specialist for a second opinion.

You should be aware that waiver processing may be time consuming, especially in complicated cases that may take additional time due to the requirement for specialist consultation and literature review.

Waiver criteria. Factors considered in granting a waiver will include the feasibility of treatment in microgravity and making provision for flight and crewmember safety. To be considered waiverable, a disqualifying condition will most likely have to comply with the criteria listed below.

a. The condition shall not pose any potential risk for subtle incapacitation that may not be detected by the individual but would affect alertness and/or ability to process information.
b. The condition shall be resolved and non-progressive.

c. The condition shall not require treatment which has the potential to impact on training and/or on-orbit operations.
d. The condition shall not pose any potential risk for sudden incapacitation.
e. The condition shall not pose any potential for jeopardizing the successful completion of a mission.

Guidelines for preparing for the medical examination

Due to the strict medical requirements for orbital certification, you will have a three-month window in which to take your medical examination, beginning three months prior to commencement of training. Those who are aware of a medical condition that may require a waiver or be disqualifying should make an appointment with a specialist and request an appropriate examination. Needless to say, individuals should not undergo a medical examination when they are ill.

Once an appointment for a medical examination has been arranged you should prepare the following:

a. Documentation detailing treatment for *any* condition(s) received in the last 5 years.
b. Documentation detailing *any* surgery performed.
c. *All* appropriate medical records pertaining to treatments, hospitalization, injury, and conditions.
d. Glasses or contact lenses if used.

The day prior to the examination you should follow the steps listed below.

a. Get a good night's sleep, be well rested, and avoid any strenuous physical activity the evening prior to or the morning of the examination.
b. Avoid the following prior to the examination:
    (i) Caffeine. Excessive amounts of caffeine may affect your electrocardiogram (ECG).
    (ii) Tobacco. Smoking prior to the examination may result in an ECG abnormality.
    (iii) Medications such as decongestants that may contain ephedrine.
    (iv) High-sugar meals. An excess amount of sugar in the bloodstream may cause an abnormal result in the urinalysis.
    (v) Eat a light meal with complex carbohydrates and proteins.

On arrival at the medical center, your reporting responsibilities include the following:

a. A report of all medicines (prescription and non-prescription) you are using.
b. A report of all visits to physicians within the last 3 years.
c. Truthfully answer all questions on the Report of Medical History Form.
d. Check all items in the Report of Medical History Form: failure to do so may delay the examination and certification.

e. Be prepared to submit additional information to the medical division of the operator.

MEDICAL STANDARDS

The purpose of these standards is to ensure that orbital spaceflight participants are physically and temperamentally fit for

a. The performance of orbital activities in weightlessness.
b. The performance of required tasks during launch and re-entry.
c. The performance of emergency egress procedures from a spacecraft.

In general, defects and diseases will be considered disqualifying for orbital spaceflight if they

a. Interfere with the performance of duties requiring dexterity, visual or auditory acuity, and the ability to speak clearly.
b. Interfere with the wearing or use of specialized equipment such as pressure suits, helmets, and controls.
c. Compromise the ability of the spaceflight participant to withstand exposure to rapid changes in atmospheric pressure, to the forces of acceleration, or to microgravity.
d. Require frequent and/or regular medical treatment or medication, or are frequently disabling.

This section describes the probable medical standards and certification procedures that orbital spaceflight participants will be required to meet. The standards are in most cases similar to those imposed for aviation service by the FAA, for spaceflight by the NASA, and for military service.

Applicability. Orbital spaceflight participant medical certification will be issued based on an individual being assessed in each of the physiological systems and conditions listed below.

1. Endocrine system.
2. Genitourinary system.
3. Respiratory system.
4. Cardiovascular system.
5. Gastrointestinal system.
6. Neurological system.
7. Psychological and psychiatric evaluation.
8. Ophthalmology.
9. Ear, nose, throat, and equilibrium.
10. Musculoskeletal system.

11. Hematological and immunologic standards.
12. General medical condition and discretionary issuance.

1 ENDOCRINE SYSTEM

You will be required to be free from metabolic, nutritional, and endocrine disorders which accredited medical conclusion indicates would render you unable to perform the duties required of a spaceflight participant under training or during flight. For an individual with diabetes mellitus, certification may be approved provided that specific control criteria are met.

Consequences for spaceflight

A potential disqualifying condition associated with the endocrine system for those hoping to fly into orbit is Type 1 diabetes mellitus, since an individual with this condition requires injections of exogenous insulin to properly metabolize carbohydrates and lipids [1]. Such a situation is clearly incompatible with orbital operations, since, in the absence of treatment, a potential exists for disastrous incapacitation [1], which may jeopardize the individual and crew.

2 GENITOURINARY SYSTEM

You will be required to be free from disease of the genitourinary tract which accredited medical conclusion indicates has the potential to render you unable to perform the duties required of a spaceflight participant under training or during flight. You will also be required to be free of any disease, medication, or surgical procedures associated with the genitourinary system that has the potential to cause subtle or sudden incapacitation.

Consequences for spaceflight

Several genitourinary conditions exist that may result in a spaceflight participant being denied certification due to the potential of such a condition to subtly or suddenly incapacitate, usually as a result of the severe pain that is often an accompanying symptom [10]. For example, an individual who suffers from urinary tract calculi is also at a significant risk of infection with hematuria, frequency, and dysuria [10].

3 RESPIRATORY SYSTEM

You will be required to be free of any significant or progressive disease of the respiratory system such as quiescent or healed lesions that are tuberculous or potentially tuberculous or have the potential to cause subtle to sudden incapacitation. You will also be required to have no established medical history or clinical diagnosis of active tuberculosis, chronic obstructive pulmonary disease, or any condition resulting from surgical procedures and/or impact trauma that has the potential for decreasing respiratory efficiency.

Consequences for spaceflight

A respiratory system and/or pulmonary function disorder increases an individual's susceptibility to dysfunction in the spacecraft environment, as such a condition carries with it a potential for incapacitation, hypoxia, acceleration atelectasis, and compromise of G tolerance. For example, an individual who suffers from chronic bronchitis and emphysema may experience significant hypoxia, a situation that will be aggravated by the hypoxic and hypobaric environment of the spacecraft. Such an individual may also suffer from dysfunction of the bronchioles which may result in small-airway occlusion in a high-G environment, such as experienced during launch. In the case of orbital passengers this problem may be exacerbated by the flight suit, which is fitted with G-protective equipment that results in a translocation of intrathoracic blood volume, a process that may further compromise small-airway function. Also, in the event of a rapid or explosive decompression (ebullism), individuals with weakened lung tissue will be at greater risk of pulmonary barotrauma.

4   CARDIOVASCULAR SYSTEM

You will be required to have no congenital or acquired abnormality of the heart that would render you unable to perform the duties required of a spaceflight participant under training or during flight. You will also be required to have no established medical history or clinical diagnosis of myocardial infarction, angina pectoris, cardiac valve replacement, permanent cardiac pacemaker implantation, or heart replacement. An individual who has coronary heart disease that is symptomatic or clinically significant on evaluation may be denied certification.

Consequences for spaceflight

The primary concern associated with the majority of cardiovascular disorders is the risk of sudden death or incapacitation [10, 22]. For example, Coronary Artery Disease (CAD) is unpredictable and may be aggravated by circumstances such as heat, hypoxia, and exposure to high $+G_z$, each of which increases myocardial oxygen demand. Another serious cardiovascular disorder, which is of particular concern for orbital passengers, is myocardial infarction, a condition associated with a range of pathophysiological behavior of atheromatous plaques which have the potential for rupture and occluding vessels, a situation that has the risk of a potentially catastrophic and incapacitating event.

5   GASTROINTESTINAL SYSTEM

You will be required to be free of disease of the gastrointestinal tract that accredited medical conclusion indicates could render you unable to perform the duties required of a spaceflight participant under training or during flight. You will also be required to be free of any surgical intervention and, in particular, any stricture or compression that has the potential to cause incapacitation.

Consequences for spaceflight

Several conditions of the gastrointestinal system may have implications for spaceflight, a number of which may be acute and/or chronic and vary in severity. For example, volumetric changes of intra-abdominal gases associated with a hypobaric environment or rapid/emergency decompression may precipitate an existing condition, and may incapacitate an individual to a degree that the individual's ability to perform an emergency egress is compromised.

   Other gastrointestinal conditions may also be incompatible with orbital operations due to their potential to incapacitate. For example, there are a number of spaceflight concerns associated with pancreatitis [10], most associated with recurring and chronic pancreatitis, the former condition requiring narcotic analgesia, and the latter resulting in extreme abdominal pain. Due to the unpredictable and often incapacitating nature of these conditions, individuals diagnosed with either recurring or chronic pancreatitis are unlikely to be considered for orbital flight.

6   NEUROLOGICAL

You will be required to have no seizure disorder, disturbance of consciousness, or neurological condition that the flight surgeon (based on case history and appropriate, qualified medical judgment relating to the condition involved) finds you unable to safely perform the duties necessary as a spaceflight participant. You will also be required to have no established medical history or clinical diagnosis of epilepsy, convulsive disorders, any history of serious head injury, alcohol or drug dependence, psychosis or neurosis, or a personality disorder that has resulted in an overt act. Other conditions that will disqualify you from flight include any disturbance of consciousness without satisfactory medical explanation of the cause or a transient loss of control of nervous system function(s) without a satisfactory medical explanation of the cause.

Consequences for spaceflight

Many neurological conditions are benign and will not present a risk to an individual during orbital flight. However, there are several conditions, such as migraines (which may develop into migrainous stroke) that have the potential to seriously compromise safety in the spacecraft environment. Similar risks exist for ischemic stroke, epilepsy, seizure, and syncope [13, 17].

7   PSYCHOLOGICAL AND PSYCHIATRIC EVALUATION

You will be required to have no established medical history or clinical diagnosis of any personality disorder that is severe enough to have repeatedly manifested itself by overt acts. You will also be required to be free of a bipolar disorder and have no history of a psychosis, or psychotic disorder, unless the psychosis was of toxic origin and there has been complete recovery. As used in this section, "psychosis" refers to a condition in which an individual has manifested delusions, hallucinations, grossly bizarre or disorganized behavior, or other commonly accepted symptoms of this condition. "Psychosis" may also refer to an individual *who may reasonably be expected* to manifest delusions, hallucinations, grossly bizarre or disorganized behavior, or other commonly accepted symptoms of this condition.

An individual who has a previous history of substance dependence will probably be disqualified from flight, except in cases where there is established clinical evidence, satisfactory to the flight surgeon, of recovery, including sustained total abstinence from the substance(s) for not less than two years preceding the commencement of flight training. As used in this section, "substance" includes the following:

a.   Alcohol, other sedatives, anxiolytics, and opioids.
b.   Central nervous system stimulants such as cocaine, amphetamines, and similarly acting sympathomimetics, hallucinogens, phencyclidine or similarly acting aryl-cyclo-hexylamines, cannabis, inhalants, and other psychoactive drugs and chemicals.

As used in this section, "substance dependence" means a condition in which a person is dependent on a substance other than tobacco or ordinary xanthine-containing (e.g., caffeine) beverages as evidenced by

a.   Increased tolerance.
b.   Manifestation of withdrawal symptoms.
c.   Impaired control of use.
d.   Continued use despite damage to physical health or impairment of social, personal, or occupational functioning.

As used in this section, "no substance abuse within the preceding 2 years" is defined as

a.   Use of a substance in a situation in which that use was physically hazardous, if there has been at any other time an instance of the use of a substance also in a situation in which that use was physically hazardous.

If your authorization or statement of demonstrated ability is denied, you will be served a letter of withdrawal, stating the reason for the action. By not later than 10 days after the service of the letter of withdrawal, you will be able to request that the flight surgeon provide for review the decision to withdraw. Your request for review may be supported by medical evidence. Within 10 days of receipt of a request for review, a written final decision will be sent to you either affirming or reversing the decision to withdraw.

Reports, falsifications, alterations, and incorrect statements

During your application process for medical certification, you will not be permitted to make a fraudulent or intentionally false statement on any application for a medical certificate. Also, if you are denied medical certification, you will not be permitted to make a fraudulent statement when requesting authorization for special issuance of a provisional medical certificate or statement of demonstrated ability. You will also not be permitted to make a fraudulent report that is made or used to show compliance with any requirement for medical certification or for any authorization or statement of demonstrated ability. Needless to say, if you reproduce for fraudulent purposes any medical certificate or alter any medical certificate, you will also be denied certification.

It is imperative that you report the truth on your medical history form, since any incorrect statement made in support of an application for provisional medical certification or request for an authorization or statement of demonstrated ability may serve as a basis for suspending or revoking your certification. Such action may also be cause for withdrawing your authorization or statement of demonstrated ability and for denying an application for a certificate or request for an authorization or a statement of demonstrated ability.

Denial of certification and medical records

If you are denied medical certification you may, within 10 days after the date of the denial, apply in writing to your operator for reconsideration of that denial. In this event, your operator's flight surgeon may request you provide medical history to support your reconsideration, or he/she may, with your permission, authorize any clinic, hospital, physician, or other person to release all available information or records pertaining to your medical history. If you fail to provide the requested medical information or history, or to authorize the release so requested, the flight surgeon may once again deny your certification.

Other roles of the Space Medical Operations Center

In addition to medically certifying you for your trip to orbit, the SMOC will also be responsible for ensuring that you are healthy prior to your flight, during your flight, and that you return to Earth in a healthy state. For example, prior to you leaving Earth, the SMOC will be responsible for the Limited Access Program (LAP), designed to protect crewmembers from exposure to contagious conditions prior to flight. The SMOC personnel will also be responsible for the preflight training of crewmembers in medical monitoring and emergency medical procedures and for indoctrinating crewmembers into the exercise training and health maintenance activities that will be required during this period. During the inflight phase, the SMOC will provide medical support for health maintenance monitoring and provide means to identify potential and unexpected health risks. The SMOC's postflight role will include means to accelerate the return of crewmembers to normal 1 G activities as soon as possible as well as providing longitudinal studies of spaceflight participants' health following their flights. The following is a chronological account of SMOC's role during the preflight, on-orbit, and postflight phases.

Eight to 10 weeks prior to launch

A flight surgeon will be assigned to your mission and will be responsible for your health and the health of your fellow crewmembers during your training. A Crew Medical Officer (CMO) will be selected during this preflight phase, who, due to the limited crewmembers, will usually be the pilot. During your training, the SMOC will deliver 15 hours of general medical training, which, in addition to general first-aid training, will cover the medical capabilities of the Flight Vehicle Medical System (FVMS). The CMO will receive the same training as you and your fellow spaceflight participants but, due to his/her role as CMO, will receive an additional 12 hours of advanced training in the fields of medical diagnostics and therapeutics and procedures.

Ten days prior to launch

At the commencement of this phase, you and your crewmembers will be placed in the LAP, which will serve as a part of the quarantine program to help protect you and the rest of the crew from contracting last-minute common medical maladies.

Four days prior to launch

Preflight medical examinations will be conducted by the flight surgeon on each crewmember.

During launch, inflight, and landing activities

The Crew Surgeon or Assistant Crew Surgeon will man the medical console at your operator's Mission Control Center (MCC), where they will serve as an integral member of the Flight Control Team (FCT). Using continuous voice and video monitoring and private daily medical conferences, your flight surgeon will be able to serve as a remote medical consultation service. If there is any medical event, physiologic data will be down-linked to the MCC and your flight surgeon will be able to evaluate the situation and provide a diagnosis. In the worst case scenario of an inflight emergency, the CMO will liaise with the Crew Surgeon to provide the necessary medical care.

After landing

Shortly after landing, you will be transported to the SMOC, where the flight surgeon will conduct postflight medical assessments and laboratory studies on you and your fellow crewmembers. The information gleaned from these tests will contribute to a medical database that will form an important part of your operator's longitudinal medical program.

Now that you understand and can prepare for the medical requirements for orbital flight, the next step is to prepare for the training phase, the details of which are covered in Chapter 6.

REFERENCES

[1] Atkinson M.A.; and Maclaran N.K. The pathogenesis of insulin dependent diabetes. *New England Journal of Medicine*, **331**, 1428–1436 (1994).

[2] Buckey Jr. J.C.; Lane, L.D.; Levine, B.D.; Watenpaugh, D.E.; Wright, S.J.; Moore, W.E.; Gaffney, A.; and Blomqvist, V.A. Orthostatic intolerance after spaceflight. *Journal of Applied Physiology*, **81** 7–18 (1996).

[3] Chi J. (Ed.). *Control of Communicable Diseases Manual*, 17th ed. American Public Health Association (2000).

[4] Christensen, J.M.; and Talbot, J.M. A review of the psychosocial aspects of space flight. *Aviation, Space and Environmental Medicine*, **57**, 203–212 (1986).

[5] Coleman, M.E.; and James, J.T. Airborne Toxic Hazards. In: Nicogossian, A.E., Huntoon, C.L., Pool, S.L. (Eds.), *Space Physiology and Medicine*, Lea & Febiger (1994), pp. 141–157.

[6] Grigoriev, A.I.; and Egorov, A.D. Physiological aspects of adaptation of human body systems during and after spaceflight. In: S.L. Bonting (Ed.), *Advances in Space Biology and Medicine*, Vol. 2. JAI Press (1992), pp. 43–82.

[7] Gundel, A.; Nalishiti, V.; Reucher, E.; Vejvoda, M.; and Zulley, J. Sleep and circadian rhythm during a short space mission. *Clinical Investigation*, **71**, 718–724 (1993).

[8] Holland, A. Psychology of spaceflight. *Journal of Human Performance in Extreme Environments*, **5**, 4–20 (2000).

[9] Kanas, N. Psychological and Interpersonal Issue in Space. *American Journal of Psychiatry*, **144**(6), 703–709 (1987).

[10] McCormick, T.J.; and Lyons T.J. Medical causes of inflight incapacitation: USAF experience 1978–1987. *Aviation Space and Environmental Medicine*, **62**, 882–887 (1991).

[11] Myasnikov, V.I.; and Zamaletdinov, I.S. Psychological states and group interactions of crew members in flight. In: C.S. Huntoon; V.V. Antipov; and A.I. Grigoriev (Eds.), *Space Biology and Medicine, Humans in Spaceflight Book*, Vol. 2. Nauka Press (1996).

[12] Nicholas, J.M. Small groups in orbit: Group interaction and crew performance on space station. *Aviation, Space and Environmental Medicine*, **58**, 1009–1013 (1987).

[13] Olshansky, B. Syncope: overview and approach to management. In: B.P. Grubb; and B. Olshansky (Eds.), *Syncope: Mechanisms and Management*. Futura (1998), p. 18.

[14] Penwell, L.W. Problems of intergroup behaviour in human spaceflight operations. *Journal of Spacecraft and Rockets*, **27**, 464–470 (1990).

[15] Pavy-Le Traon, A.; and Roussel, B. Sleep in Space. *Acta Astronautica*, **29**(12), 945–950 (1993).

[16] Polyakov, V.V.; Posokhov, S.I.; Ponomaryova, I.P.; Zhukova, O.P.; Kovrov, G.V.; and Vein, A.M. Sleep in space flight. *Aerospace Environmental Medicine*, **28**, 4–7 (1994).

[17] Rayman R.B.; Hastings, J.D.; Kruyer, W.B.; and Levy, R.A. (Eds.). *Clinical Aviation Medicine*. Castle Connolly Graduate Medical Publishers (2000).

[18] Santy P.A.; Holland A.W.; and Faulk D.M. Psychiatric diagnoses in a group of astronaut candidates. *Aviation, Space and Environmental Medicine*, **62**, 969–973 (1991).

[19] Santy, P.A.; Kapanka, H.; Davis, J.R.; and Stewart, D.F. Analysis of sleep on Shuttle missions. *Aviation, Space and Environmental Medicine*, **59**, 1094–1097 (1988).

[20] Taylor, G. Immune changes during short-duration missions. *Journal of Leukocyte Biology*, **54**(3), 202–208 (1993).

[21] Thornton, W.E.; Moore, T.; and Pool, S. Fluid Shifts in Weightlessness. *Aviation, Space and Environmental Medicine*, **58**(Suppl,), A86–A90 (1987).

[22] Wolf, P.A.; Cobb, J.L.; and D'Agostino, R.B. Pathophysiology of stroke: Epidemiology of stroke. In: H.J.M. Barnett; J.P. Mohr; B.M. Stein *et al.* (Eds.), *Stroke: Pathophysiology, Diagnosis and Management*, 2nd ed. Churchill Livingstone (1992), p. 4.

[23] *www.faa.gov/about/office_org/headquarters_offices/avs/offices/aam/ame/guide/*

[24] *www.nap.edu/info/browse.htm*

6

Training for orbital flight

At the time of writing, five civilian astronauts have visited the International Space Station, each of whom have paid $20 million for the privilege. Dennis Tito, Greg Olsen, Mark Shuttleworth, Anousheh Ansari, and Charles Simonyi each spent six months in Star City in order to complete the 900 hours training required to be qualified to fly into orbit. The training you will complete at your operator's facility will cover the same subject areas and achieve the same objectives as Dennis Tito and company, but you will spend only 248 hours spread out over only five weeks. The five weeks of training will be demanding, stressful, and challenging but ultimately rewarding, because on completion you will be trained to fly in space.

This chapter describes the theoretical and practical training you will be required to complete to qualify as an orbital spaceflight participant. Certain subjects, such as space weather, will comprise only a theoretical component, others such as survival training will comprise both a theoretical and practical element, whereas training for skills required for emergency egress will be almost exclusively practical-based. Each component of your training will be taught by subject matter experts, some of whom will be retired or current astronauts, pilots, aerospace engineers, or medical doctors.

Before applying to be trained as an orbital spaceflight participant your operator will require the following prerequisites:

1. Spaceflight participant orbital medical (see Chapter 5).
2. PADI open-water scuba qualification.
3. Basic first-aid qualification with cardiopulmonary resuscitation.
4. Payment of a 10% $500,000 deposit.

Ideally, due to the intense training, your operator will indicate qualifications that will be preferred. In certain cases, the duration of the course may be reduced, based on the spaceflight participant's previous experience. For example, if you have completed one or more of the following training components/qualifications within a 24-month period preceding your flight, you may be eligible for exemption from some of the training syllabus modules.

1. Accelerated freefall (AFF) with at least 12 jumps from 12,000 feet or above.
2. Suborbital flight experience.
3. Zero-G experience.
4. Unusual attitude flight(s).
5. Supersonic flight experience.
6. PADI open-water scuba qualification with 50 or more dives.
7. Private pilot license (PPL).
8. High-altitude indoctrination (HAI) course.
9. Sea, land, Arctic, or jungle survival.

The following represents a timeline to commencement of orbital flight training:

$T - 4$–6 months	Obtain spaceflight participant orbital medical certificate.
$T - 3$–4 months	Pay the 10% $500,000 deposit to operator.
$T - 2$ months	Receive training materials from operator.
$T - 1$ month	Pay the remaining 90% U.S.$4.5 million to operator.
$T - 1$ day	Arrive at operator training facility.

Orbital training modules

The major elements of your spaceflight participant training program include the subject areas outlined in Table 6.1. The modules are presented in this chapter in a section-by-section format and represent a basic grounding in the subject. Each module has an alphanumeric identifier that is cross-referenced within the section, the first two letters identifying the subject, the third letter identifying whether the subject is an academic or practical element and the number identifying the session number. For example, space physiology, academic module, second lesson, is identified as SPA1. An example of a typical lesson overview is provided in Table 6.2.

Assessment and qualification

In order to be qualified for orbital flight you will be required to pass exams on mission-critical subjects such as emergency procedures and administration of first aid. These exams will be preceded by an exam review the day before the exam and will be in a multiple-choice format. You will be required to score 80% to pass, and for any subject for which you score lower some remediation will be required.

To help you prepare for the lessons and exams you will have access to your operator's tutorial website. For each module of the training program the orbital training website will provide four levels of knowledge that include beginner, inter-mediate, proficient, and advanced (professional astronaut level). The first part of the

Table 6.1. Orbital training modules.

Module	Subject	Subsection(s)	Details
1	Introduction to the space environment	Space physiology Orbital dynamics Radiation and space weather	Provides an introduction to basic space physiology and exercise countermeasures, explains how rockets achieve orbit, how rocket engines work, and describes the hazards of the space radiation environment
2	Survival training	Arctic/ High-altitude Sea Desert Jungle	You will be introduced to the theoretical and practical elements of survival training in desert, Arctic, ocean, and high-altitude environments
3	Medical training	Cardiopulmonary resuscitation	Skills required to perform basic cardio-pulmonary resuscitation in microgravity
		Advanced medical intervention	Provides an overview of the use of telemedicine and ultrasonography
		Radiation monitoring	Teaches the use of the onboard radiation-monitoring systems
4	High-altitude indoctrination and G-tolerance training	High-altitude indoctrination G-tolerance training	This module is conducted at the NASTAR facility, Pennsylvania, and introduces the spaceflight participant to explosive decompression, G-tolerance training, and pressure suit operation. Also conducted at NASTAR is the 24-hour isolation mission simulation and telemedicine training
5	Incremental velocity and unusual attitude training. Zero-G flight. Sky-diving	Basic and advanced aerobatic maneuvers in a P51 Mustang	Gradually introduces the spaceflight participant to increasing velocity and exposure to unusual attitudes while performing aerobatic flight in a P51 Mustang
		Supersonic flight and basic aerobatic maneuvers	Indoctrination to Mach 1 to Mach 2 flight regime that includes increasing G maneuvers at close to Mach 1
		Zero-G indoctrination (35 parabolas)	Incremental exposure to microgravity in three phases that includes Martian, Lunar, and zero-G flight profiles
		Accelerated freefall (12 jumps)	Four practice and eight qualifying jumps from 12,500 ft to 14,000 ft

(*continued*)

Table 6.1 (*cont.*)

Module	Subject	Subsection(s)	Details
6	Onboard systems orientation and habitability training	ECLSS Operational comms GNC DAS and IVHMS	Academic and practical training in the primary systems onboard the vehicle and the habitat, to include guidance systems, environment control systems, communication systems, and crew systems.
		Food management Sleeping and hygiene General housekeeping Use of pressure suit	An overview of how to operate essential housekeeping equipment in the confines of a space vehicle/habitat. Each sub-module will comprise a theoretical and practical component
7	Prelaunch, ascent, and orbital operations. Contingency procedures	Prelaunch, ascent, orbital, and de-orbit phases	Much of this training will be conducted in a high-fidelity simulator, in which you will "fly" nominal and off-nominal missions.
		Emergency procedures	Introduction to abort options during launch, re-entry, and descent. Overview of contingency and escape systems
8	Assessment and qualification phase		Mission simulations will include ascent, orbital maneuvering, orbital rendezvous, and de-orbit and descent. During the simulation each spaceflight participant will be assessed while reacting to off-nominal situations and emergencies.

tutorial will provide an overview of the subject material, which will be followed by a 40-question test in multiple-choice format. Once you have completed the test the program will inform you of your score and provide you with feedback regarding any areas of weakness. If you score 90% you will be permitted to proceed to the next level, whereas if you score below 90% you will need to study some more before attempting the same level test. To be prepared for the exam you will need to be successful in completing the third-level tests. You will not need to attempt the tests in any particular sequence, but you should probably complete the three levels of one topic before attempting another. Once you have successfully completed the first three sections (see Modules 1–3, beginning on p. 154), the computer tutor will inform you that you are ready to attempt the real exam.

Figure 2.1. Rocketplane. Image courtesy: SpaceDev. Source: *http://www.rocketplane.com*

Figure 2.3. The Xerus spacecraft is a single-stage suborbital vehicle capable of servicing three markets: microgravity research, space tourism, and microsatellite payloads. Source: *www.xcor. com/products/index.html*

Figure 2.4. Interior of EADS Astrium's Spaceplane. Image courtesy: EADS Astrium.

Figure 2.5. EADS Astrium Spaceplane.

Figure 4.7. SpaceDev's DreamChaser^{TM} vehicle. Image courtesy: SpaceDev. Source: *www.spdv.info/index.php/photos/dreamchaser-photos/27/*

Figure 4.9. The release of t/Space's test article from a Proteus carrier plane shown in a sequence of images taken at half-second intervals. Image courtesy: t/space. Source: *www.space.com/php/multimedia/imagedisplay/img_display.php?pic = v_cxv_dropsequence_02.jpg&cap*

EADS Astrium's Spaceplane en-route to suborbit. Image courtesy: EADS Astrium.

	NASTAR, Pennsylvania					
	19	20	21	22	23	24
	Mon	Tue	Wed	Thu	Fri	Sat
13:10–14:00	HAI theory	ROR (runs 3 to 5)	Vehicle simulator orientation	Sea survival practical SSP1/ SSP2	Vehicle simulator training	
14:10–15:00	Pressure suit indoctrina-tion					
15:05–15:25	Coffee break				Coffee break	
15:30–16:20	Flight to FL80. Suit integrity checks	Emergency egress G-loading	Radiation monitoring RMP1		Unusual attitude training. P51 UAP1	
16:30–17:20			Exam review. VEG orientation			
17:30–18:20	Dinner					
19:00–20:30	CBT	Return to training facility	Exam preparation	CBT	Exam preparation	
Training hours	7 h 40 min	7 h 10 min	8 h 40 min	9 h 20 min	8 h 40 min	4 h 10 min
Total training hours, Week 4: 45 h 40 min						

WEEK 5

		Limited Access Program (Quarantine Phase)				
		25	26	27	28	29	30
		Mon	Tue	Wed	Thu	Fri	Sat
08:00–08:50	Operational comm. training COA1	Zero-G 15 parabolas (lunar) (10 CPR)	**EXAM VEHICLE SYSTEMS II.** 1 h 30 min	Equipment storage and house-keeping ESP1	Personal admin.	Preflight medical
09:00–09:50	Operational comm. training COP1			Personal admin.		
10:00–10:20	Coffee break					
10:20–11:10	Food manage-ment and preparation FMP1	Exam review. Vehicle systems II	PT session PTP5 basketball	**EXAM EMER-GENCY PROCE-DURES** 1 h 30 min	Zero-G 20 parabolas (10 CPR)	Preflight debrief
11:20–12:10	Vehicle waste manage-ment system WMP1	Sleeping quarters and hygiene SQP1				
12:15–13:05	Lunch					

		Limited Access Program (Quarantine Phase)				
		25	26	27	28	29	30
		Mon	Tue	Wed	Thu	Fri	Sat
13:10–14:00		Photography indoctrination PHP1	Simulator. Nominal takeoff and de-orbit VSP3	Simulator off-nominal takeoff and de-orbit VSP4	Unusual attitude training (supersonic T-38) UAP3	Simulator. off-nominal takeoff and de-orbit VSP5	Personal admin.
14:10–15:00		Pressurization system HPP1					
15:05-15:25		Coffee break					
15:30–16:20		**EXAM VEHICLE SYSTEMS I.** 1 h 30 min	Neutral buoyancy NB5	Radiation monitoring	Personal admin.	Simulator. Nominal/ off-nominal takeoff and de-orbit VSP6	
16:30–17:20				Exam review. Emergency procedures			
17:30-18:20		Dinner					
19:00–20:30		CBT	Exam preparation	Exam preparation	CBT	Personal admin.	
Training hours		8 h	8 h 40 min	8 h 20 min	5 h 40 min	5 h 30 min	3 h 40 min
Total training hours, Week 5: 39 h 50 min							
Total training hours: 248 h 20 min							

Module 1 Introduction to the space environment

SUBMODULE A SPACE PHYSIOLOGY

Introduction

www.atlasaerospace.net/image/podg_okp1.jpg" \t "*_blank*

Subsection A provides the spaceflight participant with an overview of the physiological consequences of exposure to microgravity and to the countermeasures required to maintain physiological conditioning and reduce orthostatic intolerance on landing.

Contents

1 INTRODUCTION

The human body has, because of its terrestrial development, adopted potent physiological mechanisms that enable man's upright posture to be compatible with Earth's 1 G environment. Microgravity adversely affects many of these mechanisms, the most seriously affected being those associated with the cardiovascular and the musculoskeletal systems. Since the effects on these physiological systems have the potential for compromising the performance of a crewmember returning from orbit, it is important that each spaceflight participant is familiar with the fundamentals of space physiology.

2   CARDIOVASCULAR SYSTEM

The cardiovascular system (CVS) consists of a heart that functions as a pump and blood vessels that function as both a high and low-pressure distribution circuit. The heart can be divided into two pumps (Figure 6.1), the chambers on the right side performing the functions of receiving blood returning from the body and pumping blood to the lungs for aeration, and the left side that performs the functions of receiving oxygenated blood from the lungs and pumping blood into the aorta for distribution throughout the body.

The low-pressure system, which is also termed the pulmonary circulation, is the pathway of blood from the right ventricle (RV) to the lungs and back to the left atrium (Figure 6.2).

The high-pressure system, which is also termed the systemic circulation, is the pathway of blood from the left ventricle to the capillaries and back to the right atrium (RA). To prevent the backflow of blood, atrioventricular (AV) valves provide a one-way flow of blood from the RA to the RV. Similarly, the semilunar valves prevent backflow into the heart between contractions.

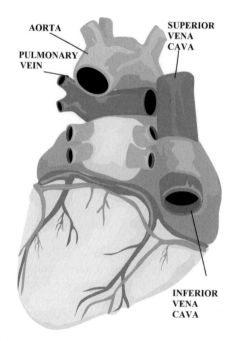

Figure 6.1. Basic heart anatomy. Image courtesy: patrimonio@BigStockPhoto.

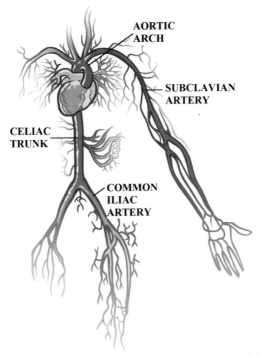

Figure 6.2. Circulatory system. Image courtesy: oguzaral@BigStockPhoto.

Blood pressure

When the left ventricle (LV) contracts, blood is forced through the aorta, which creates pressure throughout the arterial system and causes a pressure wave, or *pulse*, to travel down the aorta and throughout the arterial tree. The highest pressure generated by the heart is termed *systolic blood pressure* (SBP). Between beats, the heart pauses to allow the atria to refill with blood for the next contraction, a period of lower pressure, which is termed *diastolic blood pressure* (DBP). The difference between systolic and diastolic pressure is termed pulse pressure (PP). Normally, blood pressure is expressed as mean arterial pressure (MAP), calculated as $MAP = DP + \frac{1}{3}PP$.

Blood volume (BV)

The CVS contains 7% of the body's water in the form of plasma (about 3 liters for a 70 kg male) and serves as a major fluid transportation system, a function that has implications that are discussed later for all astronauts before, during, and following orbital flight. What is important to understand at this stage is that BV changes may occur due to changes in the water content of blood plasma, a process that is due to dynamic interaction with body tissues and blood. Blood is not passively trapped within the circulatory system because there is a constant exchange of fluid between blood plasma in the capillaries (transcapillary fluid shift) and the interstitial fluid between cells of the various tissues, an exchange that is governed by physical forces and physiological laws that explain why areas of the body can undergo dehydration or swell with excess fluid, a state termed *edema*. Edema may occur due to an increase in BP, which results in a concomitant rise in capillary pressure (P_c), which in turn causes fluid to filter out of the capillary and edema of the tissues.

2.1   Control of blood flow and blood pressure

Pressure drop, fluid flow, and resistance to that flow are the principle components of fluid mechanics. For those mathematically inclined, the control of blood flow can be described by the following equation which describes the flow of blood through a given tissue bed as being directly proportional to the pressure gradient flow across the bed and inversely proportional to the resistance encountered during transit:

$$F = (PA - PV)/R, \qquad R = 8\eta L/\pi r^4$$

where $F =$ flow
$PA =$ arterial resistance
$PV =$ venous pressure
$R =$ resistance to flow
$\eta =$ viscosity
$L =$ length of tube
$r =$ radius of tube.

Now that you understand how blood flow is regulated it is important to also understand how BP is controlled. Before putting all this together it is necessary to understand a little more about the systemic circulation. First, the central force for driving blood around the systemic circulation and through the capillaries of the organs of the body is MAP. Second, the total flow through the systemic circulation is equal to the cardiac output (CO), and third, the total vascular resistance to flow is the sum resistance offered by the entire systemic circulation, which is referred to as total peripheral resistance (TPR). Thus:

$$MAP = CO \times TPR.$$

Once again, for those of you who like equations, this relationship can be expressed as

$$resistance = 8l\eta/r^4.$$

Basically, resistance in any particular vessel will be dependent on the length of the vessel (l), the viscosity of the blood as it flows (η), and the radius to the fourth power of the vessel (r^4). The importance of this relationship is that the ability to control blood vessel radius is a very powerful tool for the body to regulate regional resistance, divert flow from one area to another, and vary BP overall. As we shall see later in this section this process is useful when it comes to dealing with the effects of microgravity.

Table 6.5. Musculoskeletal changes associated with short-duration spaceflight.

Physiologic measure	Change in microgravity
Stature	Slight increase during first week (~1.3 cm). RPB 1 day
Body mass	Postflight weight losses average 3.4%. Two-thirds due to water loss, remainder due to loss of lean body mass and fat
Body composition	Fat replacing muscle toward end of short-duration mission
Total body volume	Decreased postflight
Limb volume	Inflight leg volume decreases exponentially during first flight day. Thereafter, rate declines and plateaus within 3–5 days. Postflight decrements in leg volume up to 3%. Rapid increase immediately postflight
Muscle strength	Decreased during and postflight. RPB 1–2 weeks
Reflexes	Reflex duration decreased postflight
Bone density	Os calcis density decreased postflight. Radius and ulna show variable changes
Calcium balance	Increasing negative calcium balance inflight

5 GROUND-BASED ANALOGS

5.1 Tilt table

Head-down tilt (HDT) table testing is an analog used by gravitational physiologists to study the body's cardiovascular adaptations to changes in position and also as a technique to evaluate orthostatic hypotension. In Week 1 you will have the opportunity to experience OH following a tilt table test (TTT) and will be able to experience the sensations that occur during passive posture change, which will be similar to those you will experience in microgravity. The TTT will also be used to evaluate your response to the Stand Test, which will be conducted immediately following the tilt phase.

HDT procedure

Your TTT will take place in the morning 2 hours following a light breakfast and 6 hours following consumption of a caffeinated beverage or alcohol. On arrival at the testing suite you will be instrumented and a 15-minute familiarization tilt will be performed to acquaint you with the equipment. The practice tilt will involve 15 minutes supine rest followed by a rapid tilt upright. Ten minutes following the completion of the familiarization tilt you will be secured on the table by shoulder supports and two safety belts and tilted to a 6° HDT position. You will remain in this position for 60 minutes, during which time you will experience a cephalothoracic fluid shift of approximately one litre. During the tilt phase your BP, ECG, and HR will be monitored and you will be required to provide symptom magnitude responses on a rating scale. After 60 minutes in the HDT position you will be rapidly tilted to the vertical position, after which you will perform the Stand Test described below:

a. On completion of the HDT phase you will be assisted to the freestanding position with feet 15 cm apart.
b. You will remain in the freestanding position for 10 minutes or until presyncopal signs or symptoms appear.
c. During the Stand Test your BP, HR, and ECG will be monitored.

5.2 Bed rest

Another useful analog is bed rest, the length of which may vary from 14 days to as long as 500 days, such as the one planned by the European Space Agency (ESA) in 2007. In a recent ESA study, subjects were bed-rested for 90 days. During the 11–15-day assessment phase subjects were free to move around the bed rest facility which allowed investigators to conduct tests to determine the normal (baseline) state of subjects' skeletal, muscular, cardiovascular, endocrine, vestibular, and central nervous systems [8, 9]. Following the assessment phase, subjects spent 90 days lying in bed (except for certain tests) with their body slightly tilted downward, spending 16 hours of each day awake and the remaining 8 hours asleep. During the bed rest phase each subject was tested to determine changes in their physiological systems.

5.3 Water immersion

This analog was a favorite with the Russians who used it for several studies, some of which lasted as long as a year! To protect their subjects from maceration caused by long-term contact with water they wrapped their "volunteers" in a highly elastic waterproof fabric. Once mummified in the fabric the subject was laid in a water bath, covered with a cotton sheet, submerged to the neck, and covered up to the neck with more waterproof fabric so that only the fabric supported the subject. To ensure proper hygiene, subjects were moved from the tank for about 30 minutes each day and placed on a horizontal stretcher.

This allowed investigators to treat subjects' skin if necessary, replace sheets, and perform sanitation procedures. Subjects were fed while immersed using plastic utensils to avoid damaging the fabric and straws were supplied for drinking liquids. Bedpans were used to collect urine and feces. Reports leaked out after the Cold War suggested that some subjects committed suicide!

a.   Alpha particles. These particles have the shortest range and can travel only a few centimeters in the air and can be stopped easily by a sheet of paper or the outer layer of skin. Alpha particles are harmful only if the radioactive source is swallowed, inhaled, or absorbed into a wound.

b.   Beta particles. These particles are more penetrating, can travel several meters through the air, and can pass through a sheet of paper. Materials such as a thin sheet of aluminum foil or glass can stop them.

c.   Gamma rays. Gamma rays are electromagnetic energy with significant penetration power, requiring shielding with such materials as concrete, lead, steel, or water.

The absorbed dose of radiation is the amount of energy deposited by radiation per unit mass of material. It is measured in units of radiation-absorbed dose (rad) or in the international unit of Grays (1 Gray = 1 Gy = 1 Joule of energy per kilogram of material = 100 rad) while the milliGray (mGy) is the unit usually used to measure how much radiation the body absorbs. Because different types of radiation deposit energy in unique ways, an equivalent biological dose is used to estimate the effects of different types of radiation; this is measured in milliSieverts (mSv).

The biological effects of fast charged particles depend on the nature of the particle (its charge and velocity) and on the specific biological endpoint, such as mutation, tumor induction, or cell killing. This is expressed as relative biological effectiveness (RBE), a measure that is the ratio of the dose of gamma rays required to produce a specific effect to the dose of particle radiation required to produce the same level of effect. The RBE depends on the type of particle and the biological effect under consideration and may vary with the magnitude of the biological effect. More importantly, RBE varies greatly with the linear energy transfer (LET) of the particle, a quantity that describes the amount of energy transferred to the penetrated material per unit length. For example, high-energy protons may have an RBE value approaching 1.0, whereas high-energy iron nuclei may have an RBE value approaching 40, which means 40 times more damage is inflicted on biological tissue. Although RBE may be used as a factor to assess the damage that radiation inflicts on biological tissue, there is little information concerning the relationship between RBE and LET, although another way of expressing radiation damage is to use the radiation equivalent in man (rem), which is defined as

$$\text{dose in rem} + \text{dose in rad} \times \text{RBE}.$$

This means that 1 rem causes the same amount of biological damage regardless of the type of radiation.

Since there is little equivalent data concerning the relationship between RBE and LET, it is only possible to estimate risks to humans in spaceflight conditions. This is achieved by extrapolating from the RBE vs. LET data for cells in culture and from small mammals to humans. In addition, it is necessary to extrapolate the risks from acute exposures of humans to the low-dose-rate exposures involved in space missions (except for the relatively acute exposures from SPEs). Other factors must also be considered, such as the effects of biochemical or cellular repair reactions following exposure to HZE particles and the effects of microgravity on such reactions. But, the effects of these impacts are currently unknown [5, 9].

3   LOW EARTH ORBIT ENVIRONMENT

LEO is an orbit with a maximum altitude of 2,000 kilometers and an orbital period of about 90 minutes. An example of an LEO spacecraft is the ISS, which orbits at an altitude of 370 kilometers with an inclination of 51.6° and an orbital period of 93 minutes. The effects of radiation on humans and spacecraft systems in LEO are summarized in Table 6.7.

The LEO environment includes radiation from solar sources, Galactic Cosmic Radiation, and radiation belts (Van Allen), the latter of which consist of large toroids

Table 6.7. Low Earth orbit radiation environment.

Solar radiation	Biological effects	Satellite operations	Other systems
Extreme	High-radiation hazard for EVA astronauts	Total loss of some satellites and permanent damage to solar panels. Memory device problems may cause loss of control	No high-frequency communications possible in polar regions
Severe	Radiation hazard for EVA astronauts	Memory device problems, noise on imaging systems, interference with start trackers may cause orientation problems	Blackout of high-frequency communication and increased navigation errors
Strong	Radiation hazard avoidance required by EVA astronauts	Likely single-event problems and permanent damage to exposed components and detectors	Degraded high-frequency radio throughout polar caps and some navigation errors
Moderate	None	Infrequent, isolated problems	Small effects on high-frequency radio transmissions and navigation signals in polar regions
Minor	None	None	Minor impacts on high-frequency radio transmissions in polar regions

Geomagnetic storms	Power grids	Spacecraft operations	Other systems
Extreme	Grid system collapse and transformer damage	Extensive surface charging, orientation problems, uplink and downlink problems and loss of tracking satellites	High-frequency radio propagation impossible in many areas for 1 to 2 days. Low-frequency radio navigation disabled
Severe	Voltage stability problems. Portions of grids may collapse	Surface charging and tracking problems and orientation problems	Sporadic high-frequency radio propagation. Satellite navigation degraded
Strong	Voltage corrections required	Surface charging on satellite components. Increased drag on satellites. Orientation problems	Intermittent satellite navigation and low-frequency radio navigation problems
Moderate	High-latitude power system problems.	Corrective actions required by ground control. Changes in drag affect orbit predictions	High-frequency radio propagation fades at higher latitudes
Minor	Weak power grid fluctuations	Minor impact on satellite operations	Aurora seen at high latitudes (60°)

Source: National Oceanic and Atmospheric Administration.

of charged particles trapped in the geomagnetic field. The inner (or lower) Van Allen belt extends from some hundreds of kilometers from the Earth's surface to an altitude of 6,000 kilometers, and consists of high-energy (tens of MeV) protons and (1–10 MeV) electrons.

4   TYPES OF SPACE RADIATION

Space radiation can be divided into three types, the first of which is Galactic Cosmic Rays (GCRs). The second type of radiation that you will be concerned with while in orbit is that caused by Coronal Mass Ejections (CMEs), which cause solar protons to be emitted in large quantities. The third type of radiation is associated with Solar Particle Events, which are of particular concern to astronauts because they can deliver potentially deadly bursts of radiation over very short periods of time.

Figure 6.5. Coronal mass ejection. Image courtesy: NASA. Source: *http://antwrp.gsfc.nasa.gov.apod/ap070206.html*

Galactic cosmic rays

GCRs consist of very high–energy ions, most of which are protons (85%), alpha particles (14%), and high–atomic number ions (HZE particles ~1%) and represent the dominant source of radiation that must be dealt with onboard your spacecraft. Since GCRs possess heavy, high-energy ions of elements that have had all their electrons stripped away they are able to pass practically unimpeded through a spacecraft or an astronaut's skin [8]. Also, since these particles are affected by the Sun's magnetic field, their average intensity is highest during the period of minimum sunspot activity when the Sun's magnetic field is weakest and less able to deflect them.

Coronal mass ejections

When a coronal mass ejection (CME) occurs, large numbers of high-energy protons are released (Figure 6.5). These high-energy protons may reach LEO in less than 30 minutes, but because they are almost impossible to predict, there is little time for astronauts to prepare.

3   ESSENTIAL PRINCIPLES OF COLD-WEATHER SURVIVAL

The four fundamental principles to follow to keep warm are easily remembered by remembering the acronym C-O-L-D.

C.   Keep clothing *clean* as clothes matted with dirt quickly lose insulation value.
O.   Avoid *overheating* as clothes absorb moisture when you sweat, which in turn affects your body's ability to stay warm because the damp clothes' insulation quality is reduced. To avoid this from occurring adjust and replace clothing so the body does not sweat, and change into lighter headgear when appropriate.
L.   Wear clothing loose and in *layers*, since wearing tight clothing restricts blood circulation and increases the chance of cold injury. Tight clothing also reduces the insulating value by reducing the volume of air in the clothes.
D.   Keep clothing *dry* as inner clothing layers may become wet from sweat and outer layers may become wet from snow melted from body heat. To ensure dry clothing you should wear water-repellent outer clothing, and before entering a heated shelter make sure you brush off snow and frost.

Before moving on it is essential you are familiar with the items of your capsule survival kit, which will be issued by your operator when you commence training.

a.   Knife
b.   Waterproof and windproof matches (24)
c.   Magnetic compass (1) and world maps stored in the capsule's computer
d.   Waterproof ground cloth and cover
e.   Flashlights (2) and batteries (8)
f.   Binoculars (1) and glacier goggles (5)
g.   Emergency freeze-dried foods (15 meals) and food-gathering equipment
h.   Signalling items: flares (5), mirror (1), electronic beacon (1).

In addition, the personal survival gear stowed within the flight suit contains

a.   Utility knife
b.   Magnetic compass
c.   PDA containing topographic world maps
d.   Emergency freeze-dried foods for 24 hours
e.   Aviator™ flashlight and lithium batteries (4)
f.   Parachute flares (2).

4   MEDICAL ASPECTS

Hygiene

Washing may be impractical and will almost certainly be uncomfortable in a cold-weather environment, but it is a task that must be performed regularly to prevent skin rashes that may develop into more serious problems. To avoid this you will need to wash your feet and change your socks daily. In appropriate environmental conditions you should take a snow bath and wipe your body dry. You should change underwear at least once a week, and if you are unable to wash underwear you should at least remove it and air it. If you shave, do so before going to bed as this will give the skin a chance to recover before exposure to the cold weather.

Heat and cold regulation

Under nominal environmental conditions the body's inner core temperature remains at 37°C. The thermoregulatory system of the body allows it to react to extremes of temperature and still maintain a temperature balance, but when the body is faced with extreme cold the temperature inevitably drops, causing hypothermia, a condition caused by a lowering of the body temperature at a rate faster than the body can produce heat.

Shivering will occur at a body temperature of ≥35.5°C, a symptom that will progress to uncontrollable shivering when body temperature reaches ≤35.5°C. Sluggish thinking, irrational reasoning, and a false feeling of warmth are indicative of a body temperature between 32°C and 35°C. Progressive symptoms manifest themselves as muscle rigidity and barely detectable signs of life will occur when body temperature is between 30°C and 32°C. Finally, death is almost certain when body temperature falls below 25°C.

If you suspect a crewmember is suffering from hypothermia you should first re-warm his/her entire body by first immersing the trunk area in warm water that is between 37.7°C and 43.3°C.

A note of caution:

> *Re-warming should only be performed in a medical environment due to the increased risk of cardiac arrest and shock.*

You should then attempt to heat the body's core by administering warm water enemas, but if this is not possible the crewmember should be placed naked in a warmed sleeping bag with another naked crewmember who is also warm. You should exercise care when conducting this procedure as the assisting crewmember may also become a victim if he/she remains inside the bag too long.

When administering assistance be aware of the dangers associated with treating a hypothermia victim. If you re-warm too rapidly the victim may experience circulatory failure and ultimately heart failure. You must also be aware of "after drop", which occurs when the victim is removed from warm water, a procedure which may cause previously stagnant limb blood to relocate to the inner core.

Frostbite

Frostbite is the result of frozen tissue. *Mild frostbite* is a condition that affects only the skin which appears dull and whitish, whereas *deep frostbite* affects the skin and underlying tissues which become solid and immovable. Loss of feeling in the extremities is a signature sign of frostbite, although if loss of sensation is only for a short time the frostbite is considered mild. If loss of sensation persists and the skin becomes discolored it will be necessary to immediately re-warm the extremities by placing them next to another crewmembers stomach. A waxy discoloration of the skin indicates *severe frostbite*, a condition in which the affected body part should remain frozen as thawing will cause more damage.

To avoid frostbite you should maintain circulation in your extremities. For example, by periodically wrinkling facial skin and periodically warming your face with your hand, by wiggling your ears and by moving your hands inside your gloves.

Immersion foot

If you have been exposed to several days of wet conditions at a temperature just above freezing you may suffer *immersion foot*, a condition that in its mildest form is characterized by pins and needles, tingling, numbness, and pain. In its moderate form your skin will become red then bluish and finally black and your feet will become cold and swollen. At this point you will find it difficult and painful to walk. Finally, in the condition's advanced stage your muscle and nerve tissue will necrotize and amputation will become necessary.

Dehydration

The material of your flight suit absorbs moisture that evaporates in the air, which means it is essential that you hydrate. To determine hydration levels check the color of your urine on the snow. If it is dark you are dehydrated, whereas if your urine is a light yellow color you are optimally hydrated.

Snow blindness

Ultraviolet radiation glare from the snow may cause snow blindness, the symptoms of which include a sensation of grit in the eyes and pain in and over the eyes. Treatment will require you to bandage both eyes until symptoms disappear. To avoid this condition make sure you wear glacier glasses or cut slits in appropriate material to reduce glare and slow the onset of symptoms.

5   SHELTERS, FIRE, WATER, AND FOOD

Shelters

Regardless of which type of shelter is built, certain basic principles apply. Always ventilate enclosed shelters, always block a shelter's entrance to keep wind and snow out, never sleep directly on the ground and never sleep without extinguishing the stove.

Fire

Your ability to make fire will largely be determined by whether you landed within or outside the tree line. If fortunate enough to land in a subarctic region you will be able to use driftwood, willow, and alder as fuel sources, whereas if you land within the tree line you will also be able to use the tamarack tree, a useful tree to burn since it makes a lot of smoke. Regardless of where you land you will need to leave the fuel in the tanks for storage, bearing in mind not to expose flesh to petroleum or oil in extremely cold temperatures as contact may cause frostbite. If building a snow shelter be wary of excessive heat as this will melt the insulating layer of snow and any fire inside a shelter lacking adequate ventilation may result in carbon monoxide poisoning.

Water

Although Arctic and subarctic water sources are generally more sanitary than other regions, always purify water prior to drinking. Before ingesting, be sure to completely melt snow and ice since melting either in the mouth may cause internal injuries and will also cause loss of body heat. If pack ice is available it is possible to use sea ice to melt for water as this loses its salinity over time. Body heat may also be used to melt snow by placing snow in a water bag and placing the bag between layers of clothing. When both ice and snow are available, melt ice rather than snow as the water yield from a fixed volume of ice is greater than for the same volume of snow.

Food

The generic emergency ration pack is designed to provide 2,000 calories per 24-hour period for 3 days. If, after this time you have not been recovered you will need to seek alternative food sources which will be determined by your location and time of year. If you land near the coast then obvious supplementary food sources will include kelp and seaweed and herring gull eggs. If you have landed in the far north you should be wary of the black mollusk which may be poisonous due to toxins produced in the mussel's tissue.

Arctic and subarctic wildlife

In the Arctic coastal region you may be unlucky enough to encounter polar bears. Ideally, you should avoid polar bears but if it is necessary to kill one for food or self-defence, the point of aim should be the brain. Always cook polar bear meat before eating and never eat polar bear liver due to the toxic concentrations of Vitamin A. In this area you may also encounter seals, which in spring make good targets as they spend much of their time basking on the ice beside their breathing holes, which also serve as their means of escape if they happen to encounter their primary enemy, the polar bear! Approach a seal downwind while it sleeps, and if it moves imitate its movements. Approach should be made with your body sideways to the seal and your arms close to your body so you look as much like a seal as possible. When within 20 m to 30 m of the seal, kill it instantly, and recover the seal quickly before it slides into the breathing hole. If you are fortunate to have landed south of the tree line you may be able to catch owls, ptarmigans, jays, grouse, or ravens. Rock ptarmigans normally travel in pairs and can be easily approached, whereas willow ptarmigans gather in large flocks and can be easily snared. Also be aware that the Arctic and subarctic support a variety of fauna in the form of edible plants during the warmer months, such as shrubs and reindeer moss.

6   TRAVEL

If, after a couple of days you have not been rescued, or if you need to reach a place in order to be rescued, there are some basic principles that you should follow when travelling in subarctic and Arctic regions. Always travel in the early morning when temperatures are coldest if you are in an area prone to avalanches. This time of day is also the best time to cross streams as they are most likely to be still frozen. Other weather factors that may affect your progress include any weather that may reduce your ability to see, so avoid traveling in whiteout or blizzard conditions. Inevitably, sooner or later you will probably encounter areas of crevasses and snow bridges. The former you should try to avoid and the latter you should cross at right angles to the obstacle it crosses.

DESERT SURVIVAL

Contents

1   THE DESERT ENVIRONMENT

A desert is any environment that receives an annual rainfall of less than 25 cm, an evaporation rate that exceeds precipitation, and a high average temperature. There are 50 major deserts in the world that meet this definition, but despite the conventional images of sand dunes, deserts are not necessarily oceans of sand. Much of the Sahara and the Kalahari, for example, consists of rocky plateaus (*hammadas*), whereas large wastelands (*takyrs*) make up the deserts of Soviet Asia. It is important you are familiar with the characteristics and nature of the terrain on which you have landed since this will be essential in determining the best course of action you take to stay alive.

2   WEATHER

Temperature

Although you can expect very high temperatures during the day, which may approach 55°C in the shade, also be prepared for the very low temperatures during the night, which may in certain deserts, such as the northern Gobi, fall to −40°C.

Precipitation

Some deserts such as the Atacama have experienced no rainfall for over 200 years whereas some mountainous deserts may receive as much as 20 cm of annual rainfall. Rain, when it does fall, can be hazardous since it often falls in the form of cloudbursts that drop huge amounts of water onto a baked earth, a condition that often results in flash floods strong enough to carry people away.

Winds

A typical desert is often characterized by vast expanses devoid of prominent geological formations. Winds that move across such a topography acquire strong speeds and steady direction that routinely reconfigure the landscape, making it difficult to navigate. The winds are often blisteringly hot, which will cause overheating and will also interfere with radio communication.

3   WILDLIFE

Depending on the type of desert you find yourself in you can expect to encounter snakes, spiders, lizards, and/or scorpions. Some of these creatures are harmless, whereas others such as the Mojave rattlesnake, the black widow spider, and the bark scorpion should be avoided.

4   SURVIVAL EQUIPMENT

Desert survival equipment is contained in the stowage compartment behind the pilot's seat and contains the following:

a.   Short machete (1) and Gerber knife (1)
b.   Waterproof and windproof matches (24)
c.   Brunton™ 8099 Eclipse magnetic mirror sight compass (1)
d.   World maps stored in the capsule's computer

Table 6.14. The fall in body temperature with corresponding symptoms experienced at each temperature.

Body temp (°F)	Signs and symptoms
99	Shivering
97	Impairment of manual dexterity
95	Errors of commission or omission
93	Muscle function impaired
91	Introversion
89	Slowing of mental and physical activity
87	Amnesia. Pupils dilating
86	Shivering replaced by muscle spasticity
84	Unconsciousness
82	Ventricular fibrillation likely
79	Slowing of respiration and heart rate
77	Muscle flaccidity
76	Death

Hypothermia occurs rapidly due to the decreased insulating quality of wet clothing and water displacing the layer of still air that surrounds the body. To reduce the chances of hypothermia the following guidelines should be followed:

a. Each crewmember should don an anti-exposure suit.
b. Rig a spray shield and canopy.
c. Cover the raft floor with canvas for extra insulation.
d. Huddle to maintain warmth and move to maintain blood circulation.

Hot-weather considerations

Humans can tolerate a drop in core body temperature of 10°C—but an increase of only 5°C. Exposure to excessive heat may result in thermal stress conditions such as heatstroke, the most serious of the heat -stress maladies. To reduce the chances of suffering from thermal stress, the following guidelines should be followed:

a. Rig a canopy to ensure sufficient space for ventilation.
b. Cover skin to protect from sunburn.
c. Use sun block on all exposed skin.
d. Drink little and often.

5   SHORT WATER AND FISH AND FOOD PROCUREMENT PROCEDURES

"After a week, the terrible thirst became a bigger problem than the general discomfort and intense heat from the sun. It was no longer simply a question of a dry mouth; now our tongues were swollen and furred, while our lips were cracked. It was difficult to muster a spit and eating our tack was impossible. After a quarter of an hour of chewing we still couldn't swallow it and in the end simply blew the powder away like dust."

World War II ship survivor

If water supply is limited and it is not possible to replace it by chemical or mechanical means the guidelines described here should be indicated.

a.   Protect freshwater from seawater contamination.
b.   Ensure the body is shaded from the Sun and reflection off the sea and dampen clothes during the hottest part of the day.
c.   Limit exertion to a minimum.
d.   Determine the daily water ration based on the following:
   •   Amount of water remaining
   •   Solar still's output and desalting equipment
   •   The physical condition of each crewmember.
e.   Eating when nauseated should be avoided.
f.   To minimize water loss through sweating, soak clothes in the sea and wring them out before donning.
g.   Be prepared for showers/rain by keeping the tarpaulin ready for collecting water.
h.   At night, secure the tarpaulin and turn up its edges to collect dew.

Water from fish

Drink the aqueous fluid found along the spine and in the eyes of large fish by carefully cutting the fish in half to obtain the spinal fluid. If water is so short that it is necessary to drink fish fluids, do not drink any other body fluids as these are rich in protein and fat and will use more of your reserve water in digestion that they provide.

Sea ice

In Arctic waters it is possible to obtain water from old sea ice, which is easily recognizable by its bluish hue and rounded corners. Old sea ice is practically free of salt and is therefore safer than new ice, which is grey and salty.

Food procurement

In the open sea, fish will the primary food source. There are poisonous and dangerous fish, the details of which will be provided by your instructor, but in general, when out of sight of land, fish are safe to eat.

Fish

When procuring fish the following guidelines may prove helpful:

a. Do not handle the fishing line with bare hands or tie the line to the raft as salt adhering to it can make it a sharp cutting edge that is dangerous to the raft and your hands. Before handling the line wrap an item of clothing around your hands to protect them.

b. In warm regions, gut and bleed fish immediately after catching them. Cut fish that you do not eat immediately into thin strips and hang to dry. Fish not cleaned and dried may spoil in a matter of hours, especially fish with dark meat. Any fish not eaten should be kept for bait.

c. Never eat fish that have pale, shiny gills, sunken eyes, flabby skin and flesh, or an unpleasant odor.

d. Do not confuse eels with sea snakes that have an obviously scaly body and a strongly compressed, paddle-shaped tail. Both eels and sea snakes are edible, but must be handled with care due to their poisonous bites. The heart, blood, intestinal wall, and liver of most fish are edible.

e. Shark meat is good but spoils rapidly due to high concentrations of urea in the blood. To prevent spoiling, the shark should be bled immediately and the meat soaked in several changes of water. Consider all shark species edible except the Greenland shark whose flesh contains high quantities of Vitamin A. Shark livers should not be eaten due to excessive Vitamin A content.

Fishing aids

The following materials may be used to make fishing aids:

a. Fishing line. Unravel threads from the tarpaulin and tie them together in short lengths in groups of three. Parachute suspension cord also works well.

b. Fish lures can be fashioned by attaching a double hook to any shiny piece of metal.

c. Grapples may be used to hook seaweed, which may then be shaken to obtain crabs and shrimp. These may then be either eaten or used for bait.

d. Bait. Use the guts from birds and fish for bait.

6 MEDICAL PROBLEMS ASSOCIATED WITH SEA SURVIVAL

Seasickness

The nausea and vomiting caused by the motion of the raft may result in extreme fluid loss and exhaustion, crewmembers becoming seasick, attraction of sharks to the raft, and unsanitary conditions. It is important therefore to treat seasickness by washing the crewmember suffering from the condition and washing the raft to remove the sight and odor of vomit. The affected crewmember should lie down and rest, be administered medication, and be prevented from eating food until nausea has gone.

Saltwater sores

These sores result from a break in skin exposed to seawater for an extended period and may form scabs and pus. They should be treated by being left closed and flushed with fresh water, if available, and allowed to dry, after which antiseptic should be applied.

Constipation

Laxatives should not be administered as this will cause dehydration. To prevent this condition, adequate amounts of water should be taken and exercise should be performed as far as is possible within the confines of the raft.

Sunburn

This can be prevented by remaining in the shade, keeping the head and skin covered and using high-factor blocking. To avoid reflection from the water the canopy should be erected to optimize shade.

Heat illness

This condition is associated with general malaise caused by a rise in core body temperature from other than pathological causes such as infections and occurs when heat gain by the body exceeds heat loss. In this condition body temperature rises and will continue to do so until some alleviating measure re-establishes heat balance. The progression of the signs and symptoms of the condition are detailed in Table 6.15.

Table 6.15. Signs and symptoms of heat illness.

Heat exhaustion	Chronic effects of heat
Light-headedness, dizziness, faintness	Lassitude
Rapid, shallow breathing	Discomfort
Rapid, thready pulse	Irritability
Pins and needles of fingertips	Appetite suppression
Feeling very hot	Impaired physical and mental performance
Hot, flushed skin; sweating	Muscle cramps
Nausea and vomiting Visual disturbances and headache	Heatstroke, often characterized by hot, dry skin leading to unconsciousness, brain damage, and death

Dehydration

The simple remedy for this is to administer oral re-hydrating fluid, which should include half a teaspoon of salt, five level teaspoons of sugar mixed with a liter of water.

Non-Freezing Cold Injury (NFCI)

Tissue temperatures between 17°C and −0.5°C that persist for a protracted period may result in an NFCI. In a typical NFCI case the exposed limb will feel uncomfortable before becoming numb and suffer impaired function. As the victim re-warms, circulation will normally return, except in the severest cases, and will be accompanied by tingling. Painkillers will usually be required to alleviate the pain of re-warming.

Freezing Cold Injury (FCI)

The ambient temperature must be below freezing or significantly lower for this condition to occur. This may be the case when a crewmember is immersed in seawater that is close to freezing. The initial treatment of an FCI-affected crewmember requires removal of clothing and whole-body warming in agitated warm water. If re-warming is not possible the victim should be administered painkillers and the affected limb(s) warmed slowly.

7 SHARKS

Only about 20 species of shark are known to attack humans, the most dangerous of which are the great white, hammerhead, mako, and tiger shark. Other dangerous sharks include the gray, blue, sand, nurse, bull, and oceanic white-tip sharks. Generally, sharks in tropical/subtropical seas are far more aggressive than those in temperate waters.

Sight, smell, sound, and vibrations in the water guide sharks to their prey. So sensitive is this latter sense that even a fish struggling on a line will be sufficient to attract a shark's attention.

Although most reported attacks occur during late afternoon, sharks feed at all hours of the day. To reduce the chance of shark attack the following guidelines may prove useful:

a. Stay with other crewmembers.
b. Urinate in small amounts only and allow it to dissipate between discharges.
c. If a shark attack is imminent, splash and yell, to keep the shark at bay. Yelling underwater or slapping the surface may also scare the shark.
d. If attacked, kick and strike the shark, aiming for the gills or eyes, but *not* the nose. If you hit the nose, you may injure your hand if it glances off and hits the shark's teeth.

If sharks are sighted, cease fishing, and if you have hooked a fish, let it go. Do not throw garbage overboard and absolutely do not let your arms, legs, or equipment hang in the water! Instead, focus on keeping quiet and not moving.

8   DETECTING LAND; RAFTING AND BEACHING TECHNIQUES

Detecting land

The following are indicators that land is near:

a.   A fixed cumulus cloud in a clear sky, often hovering over or slightly downwind from an island.
b.   In the tropics, the reflection of sunlight from shallow lagoons or shelves of coral reefs often causes a greenish tint in the sky.
c.   In the Arctic, light-colored reflections on clouds often indicate ice fields or snow-covered land.
d.   Deep water is dark green or dark blue. Lighter color indicates shallow water, which may indicate proximity to land.
e.   There are usually more birds near land than over open ocean. The direction from which the flocks fly at dawn and to which they fly at dusk may indicate the direction to land.
f.   Mirages occur at any latitude, but are more common in the tropics, especially during midday.

Rafting and beaching techniques

Once land has been sighted a landing should not be attempted when the Sun is low and ahead of the raft. Instead, attempt a landing on the lee side of an island by aiming for gaps in the surf line and avoiding coral reefs, rocky cliffs, and rip currents. If it is necessary to travel through surf, keep clothes and shoes on to avoid cuts and inflate life vests. The sea anchor should be trailed over the stem using as much line as necessary. If there is a strong wind and heavy surf, the raft must have all possible speed to pass rapidly through the oncoming crest to avoid being turned broadside. If in a medium surf with no wind or offshore wind, prevent the raft from passing over a wave so rapidly that it drops suddenly after topping the crest. As the raft nears the beach, ride it on the crest of a large wave. Paddle or row as hard as possible and ride in to the beach as far as you can. Do not jump out of the raft until it has grounded, then quickly disembark.

9   SWIMMING ASHORE, PICKUP, AND RESCUE

Swimming ashore

If it is necessary to swim ashore, wear shoes and at least one layer of clothing. If the surf is moderate, ride in on the back of a small wave by swimming forward with it and then dive to a shallow depth just before the wave breaks. In high surf, swim toward shore in the trough between waves, then, when the seaward wave approaches, face it and submerge. After it passes, work toward the shore in the next trough. If you must land on a rocky shore, search for a place where the waves rush up onto the rocks and avoid places where waves explode with a white spray. After selecting a landing point, advance behind a large wave into the breakers, face toward shore, and take a sitting position with your feet in front, 2 or 3 feet lower than your head, a position that will allow your feet to absorb the shock when you land.

Pickup and rescue

On sighting a rescue craft all lines and other equipment that may cause entanglement should be cleared. All loose items in the raft should be secured, and canopies and sails should be struck. Crewmembers should fully inflate life preservers, remove all other equipment, and remain in the raft unless otherwise instructed.

TROPICAL SURVIVAL

Contents

1 TROPICAL CLIMATES

There are four types of tropical climates: rain forests, primary and secondary jungle, tropical savannas, and deciduous forests. Most are rich in nutritious vegetation, sources of water, and possibilities for shelter.

Rain forests

Typical rain forests are characterized by a hot, steamy climate with mean monthly temperatures between 24°C and 28°C and are found in Central and South America, Indonesia, Southeast Asia, West and Central Africa, and tropical Australia. A feature of this type of tropical forest are the three distinct layers of tree crowns, the highest of which forms the forest canopy, the layer below composed of young trees, and the lower level consisting mainly of tree branches and foliage.

Primary and secondary jungle

Primary jungle is strongly tiered and is characterized by tall trees and layers of vegetation below, whereas secondary jungle is less tiered and is the result of the clearing of primary jungle for cultivation.

Tropical savanna

Savanna is found in tropical regions 8° to 20° from the equator and receive a mean annual precipitation between 80 cm and 150 cm, most of which falls between April and September in the northern hemisphere, and between October and March in the southern.

Deciduous forest

This may also be called monsoon forest and is characterized by teak trees that shed their leaves in the dry season, and bamboo thickets.

2   WEATHER

There are two rainy seasons which correspond with the equinoxes, and apart from these periods, which may last several weeks, the weather is usually sunny. The very high humidity never varies in those areas with dense canopies, and despite a cooler maximum temperature than a desert, the conditions in most tropical environments are harder on the body because such high humidity prevents thermal regulation.

3   WATER

Microbes, parasites, and viruses may contaminate many of the water sources, and it should be assumed that all jungle water other than from bamboo and direct rainfall is contaminated.

Water sources

a. Plant sources of water may be obtained from the banana and plantain trees by cutting the trees down and sawing through the trunk. The stump should then be hollowed out to form a bowl, which will then fill with water. Bamboo that contains water can be recognized by its sharp 45° inclination in relation to the ground and its yellow–green color. It is possible to actually hear the water inside the bamboo as it is shaken. Bamboo makes for an ideal source of fresh and cool water since, even if the temperature of the environment is very high, it will always maintain a low temperature. Maloukba is a large water-bearing tree found in Southeast Asia and New Guinea, but since the water does not flow during the day, it is necessary to wait until after sunset before making two cuts in the form of a "V" to collect up to 150 liters of water. Lianas and vines can be cut to obtain water by cutting deeply into the vine as high as possible and then severing it completely near the ground.

b. Streams. If the stream is fast-flowing with a stone and sand bed, the water is likely to be pure, although as a precaution the water should still be purified in case there are animal deposits upstream.

c. Collecting dew. By tying cloth around your ankles and walking through dew-covered grass before sunrise the cloth will become impregnated with water which can then be wrung out.

4 FOOD SOURCES

Wildlife

Due to the dense vegetation and restricted visibility the most common type of wildlife encountered will probably be reptiles and insects although you will receive training in setting animal traps and baiting in case you encounter other food sources such as those listed below:

a. Snakes are best eaten either boiled or fried. To prepare for eating, the snake's head should be removed and the stomach skin slit downward from the neck and peeled back until completely removed.

b. Monkeys. Due to the aggressive nature of larger males, hunting should be restricted to the smaller species, the signs of activity of which may be found near watering holes and tracks on trails.

c. Tapir are pig-like herbivores found in Sumatra, Central and South America, and Malaysia. Tapirs like to sleep by day and feed by night, and are often found near swamp areas.

d. Wild pigs are found in all types of rain forest, and usually live in groups. As these animals are highly habitual they are relatively easy to hunt, although caution should be exercised when preparing its meat as it is often infested with worms!

Plants

It is important to be able to distinguish between those plants that are edible and those that are poisonous if you are to survive in the jungle. Your instructors will provide you with a chart to aid identification during your training, information which is also provided in your PDA. An exhaustive listing of plants is beyond the scope of this section, but the following are some of the most commonly found edible and non-edible plants:

Edible plants

Bamboo, which is found in the Far East, may be eaten raw although the fine black hairs on the edge of the leaves are poisonous. Bananas are found in the humid tropics and are characterized by large leaves and flowers that hang in clusters. The breadfruit tree is found in the South Pacific, West Indies, and Polynesia and is recognized by its dark green leaves. Manioc, also known as tapioca or cassava, is found in the tropics and is easily recognized by its large, tuber-like roots, although the bitter-tasting type of manioc contains poisonous hydrocyanic acid and is best avoided. The taro plant is found extensively in the tropics wherever there is moist ground and is characterized by large heart-shaped leaves, orange flowers, and a large turnip-like tuber. Yams, of which there are more than 700 species, are found extensively in the tropics and can be recognized by their enormous tubers. Yams must be cooked as they are poisonous if eaten raw.

Inedible plants

Castor beans, found in tropical Africa, grow to 12 m in height and are distinguished by orange flowers devoid of petals. The bean-shaped seeds contain castor oil and are deadly poisonous. Hemlock, the distinguishing features of which include an unpleasant smell and hollow purple-spotted stems, is highly poisonous and may be fatal if ingested. Manchineel, which is found in the southern U.S., Central America, and northern South America may grow to 15 m, and is characterized by shiny green leaves and small greenish flowers. The fruit is poisonous. The physic nut, a shrub found in the southern U.S. and throughout the tropics has small, green/yellow flowers and an apple-sized fruit which is poisonous. Unsurprisingly, given its name, the strychnine tree, which is found in Australia and Southeast Asia has greenish flowers and orange berries that are poisonous.

5   NAVIGATION AND TRAVEL

In the unlikely event that the landing site is visible from the air and signaling devices are available, the best course of action may be to stay in place and wait for aerial searches. If chances of location are judged to be low, jungle travel may be the best option assuming minimal casualties.

The following are some guidelines for navigation through the jungle:

a. Tracks that are obviously made by animals will usually lead to water, whereas human-made ones will usually lead to villages.
b. Dense vegetation will make it impossible to travel in a straight line and will usually reduce progress to less than 500 m/h. To increase speed, geographical features such as ridge lines should be used, although rivers should be avoided due to their meandering courses which may increase linear distance threefold.
c. Mangrove swamps that occur on coastlines should be avoided or crossed as quickly as possible as they have extensive, tangled root systems above and below water that may harbor crocodiles and leeches.
d. Savanna should be avoided, due to its thick broad-bladed, sharp-edged grass that stands between 1 m and 5 m high, thus reducing visibility and providing little shade from the Sun.

Travel through the jungle is akin to negotiating an obstacle course, but there are guidelines that make travel easier.

a.   Choose routes that follow the largest opening in vegetation.
b.   Move slowly and purposefully, stopping regularly to check the GPS.
c.   Do not attempt travel after sunset.
d.   Do not try and cut a path to follow a compass bearing.
e.   Take precautions to limit the effect of insects and leeches.
f.   Do not swim unless you are certain you can negotiate the current.

6   HAZARDS

Snakes and insects can make the jungle a living hell. For example, there is a species of assassin bug (*Reduvioidea*) that transmits sleeping sickness, and there are certain mosquitoes capable of transmitting a cocktail of diseases, such as malaria, yellow fever, and filariasis. There are red ants capable of delivering several nasty bites, flukes and hookworms that enter the body by piercing the skin before moving into the bloodstream, and the anopheles mosquito, a deadly mosquito that carries malaria. Your instructors will provide you with a more thorough indoctrination to these and other hazards.

7   MEDICAL PROBLEMS

During astronaut training each crewmember will be vaccinated against the following:

a.   Cholera
b.   Diphtheria
c.   Yellow fever
d.   Hepatitis A and B
e.   Japanese B-encephalitis
f.   Meningococcal meningitis
g.   Poliomyelitis (IPV)
h.   Poliomyelitis (OPV)
i.   Rabies
j.   Tetanus
k.   Tick-borne encephalitis
l.   Tuberculosis
m. Typhoid fever.

Despite all these vaccinations, there are many tropical diseases, infections, and medical dangers that you may be exposed to, but there are also several ways of avoiding such hazards some of which are described below:

a. Foodborne diseases such as ciguatera, tetrodotoxin, and paralytic shellfish poisoning result in severe neurological symptoms and vomiting, but can be prevented by not eating certain fish.

b. Poor hygiene encourages fungal infections and can be avoided by washing frequently.

c. Amoebiasis occurs in tropical and sub-tropical areas, the causative agent being a pathogenic protozoa that can eventually infest the lumen of the colon and the bowel wall. The infection is acquired by fecal–oral transmission and can be avoided by implementing good hygiene.

d. Giardiasis is caused by protozoa transmitted via the oral–fecal route, by smear infection, or from contaminated food. Symptoms include heavy diarrhoea, mal-absorption, and abdominal pain. Again, effective hygiene measures can prevent this being a problem.

e. Schistosomiasis is an infection characterized by fever followed by hematuria and bloody stools. It can be avoided by not swimming or wading in lakes and rivers.

f. Borreliosis is a nasty disease transmitted by ticks which results in painful skin, joint, cardiac, and neurological symptoms. The possibility of infection can be reduced by effective use of insect repellent and covering of exposed skin.

Table 6.20. Advanced Life Support Kit.

Chest drain	Tape	Gauze
Catheters	Syringe (10 cc)	Alcohol wipes
Oral and nasal airways	Scalpel	Tracheostomy tube
Nasogastric tube	Surgical gloves	Laryngoscope
Magill forceps	Proventil inhaler	Curved scissors
Endotracheal tube	Xylocaine jelly	

Assessment kit			
Pulse oximeter	Oral thermometer	Penlight	Tongue depressors

Emergency surgery contents			
Forceps	Bandage scissors	Scalpels	Steri-strips
Needle driver	Sutures	Hemostats	Sterile drape

Additional contents		
IV infusion pump	Automated blood pressure cuff	Diagnostic algorithms

ALSK drug contents		
Adenocard	Epinephrine (1:10,000)	Morphine
Atropine	Furosemide	Narcan
Dexamethasone	Haldol	Nitroglycerine
Diazapam	Inderal	Romazicon
Dopamine	Lidocaine	Verapamil

Medical checklist

The Basic Spacecraft Medical Kit (BSMK) includes a manifest of pharmacologic interventions and various diagnostic items, such as a stethoscope and thermometer. Although you will become familiar with the use of many of the diagnostic items during your training, you will not be required to have a working knowledge of drug administration, as most pharmacologic interventions can be performed using the appropriate algorithm stored in your personal data assistant (PDA).

Advanced Life Support Kit (ALSK)

In addition to the items found in the BSMK you will also be trained in the use of some of the instruments contained in the Advanced Life Support Kit (ALSK). The ALSK includes medical instruments and supplies designed to support Advanced Cardiac Life Support (ACLS) and Basic Trauma Life Support (BTLS), procedures that you will have the opportunity to practice on the ground, during parabolic flight, and during the telemedicine exercise that is conducted during the isolation mission in Week 3.

Diagnostic algorithms such as the one in Figure 6.6 are contained in a file in your PDA and also in a medical-specific PDA located in the ADSK for myocardial infarction, asystole, post-resuscitation, breathing difficulty, tachycardia, crycothyrotomy, injections, shock, vertigo, barotraumas, shock, and burns.

BAROTRAUMA - EAR BLOCK/SINUS BLOCK
(ISS MED/3A - ALL/FIN)   Page 1 of 1 page

BAROTRAUMA - EAR BLOCK, SINUS BLOCK

NOTE
Symptoms result from reduction in barometric pressure and expansion of trapped gas. Symptoms may occur during decompression preceding EVA or following loss of cabin pressure. Pain should resolve after repress in most cases. Persistent ear pain following repress requires examination.

Symptoms
Abdominal distention
Ear pain
Inability to clear ear
Loss of hearing acuity
Sinus pain
Toothache
Jaw pain

Treatment
AMP   1. If ear pain persists following repress, perform Otoscope Exam (Physical
(blue)   Exam-9).
              Look for the following signs:
                Red, inflamed ear drum
                Perforation of eardrum
                Drainage from ear drum, clear or bloody

          2. Contact Surgeon with results.

Figure 6.6. Diagnostic algorithm.

2 PRINCIPLES AND METHODS FOR RENDERING FIRST AID IN MICROGRAVITY

Since one of the prerequisites for this course is a basic life support qualification, there is no requirement to describe procedures in detail, although prior to discussing these procedures in a microgravity environment it is worthwhile to review the actions required in administering aid.

Cardiopulmonary resuscitation

The essential principles of administering cardiopulmonary resuscitation (CPR) are reviewed in Table 6.21. In unconscious or collapsed individuals, the state of ventilation and circulation must be determined immediately. Speed, efficiency, and proper application of CPR directly relate to successful neurological outcome, as tissue anoxia for more than 4 to 6 minutes can result in irreversible brain damage or death. Successful CPR depends on early BLS, prompt recognition and treatment, and airway and rhythm control as necessary and must be continued until the cardiopulmonary system is stabilized, the individual is pronounced dead, or resuscitation cannot be continued.

2.1 Cardiopulmonary techniques in microgravity

CPR in microgravity remains a challenge that has been addressed in several investigations performed onboard Space Shuttle missions and parabolic flights.

Table 6.21. The ABCDs of cardiopulmonary resuscitation.

A	*Airway opened* Establish airway patency using head tilt chin lift, head tilt–neck lift, or mandibular jaw thrust If available, use an artificial airway in the unconscious patient
B	*Breathing restored* Note chest movement If no spontaneous chest movement, initiate mouth-to-mouth rescue breathing Reassess for chest movement
C	*Circulation restored* Establish pulselessness If pulseless, rapidly assess for VF or pulseless VT If VF or pulseless VT is present, defibrillate If pulseless and appropriate equipment is not available, begin chest compressions
D	*Defibrillate*

VF = ventricular fibrillation; VT = ventricular tachycardia.

During your medical training phase you will be trained in the following four CPR techniques:

1. Side straddle (STD). Two rescuers delivering synchronized chest compressions and ventilations from the victim's right side.
2. Waist straddle (WS). Two rescuers, in which one rescuer performs compressions by kneeling across the victim's waist.
3. Handstand position (HS). In which one rescuer places their feet on the flight deck ceiling, providing chest compressions through quadriceps muscle group extension.
4. Reverse bear hug (RBH). A modified Heimlich maneuver where one rescuer performs chest compression from behind the victim.

You will have the opportunity to practice these procedures during the parabolic flight component of your training when you will be accompanied by two flight surgeons who will assist you in performing the ALS procedures. Once again, you will be using the HPS, which will be fastened to a table. During the first five parabolas you will practice basic resuscitation techniques. During parabolas 6 to 10 you will attempt intubating the HPS with an endotracheal tube, which will be challenging since you will only have 20 seconds in which to accomplish the task!

To assist you, the HPS will be strapped into the supine position 25 cm from the flight deck and restrained using the CMRS using anchoring tracks. Each crew-member will perform each CPR position over four parabolas with an interval of one parabola between positions. Physiological-equivalent data will be collected for the purposes of feedback during the postflight briefing; this will include the following parameters:

a. Compression rate and depth
b. Endotracheal tube pressure change per unit time
c. Tidal ventilation (TV) and minute volume (MV).

2.2   Resuscitating patients in ventricular fibrillation

Ventricular fibrillation (VF) refers to a state in which the muscle fibers of the ventricular myocardium are contracting in a highly and unsynchronized way. The most important technique for resuscitating patients in VF is defibrillation, a proce-dure that is made simple using the Lifepak defibrillator that will guide you through the steps using screen messages, voice prompts, and lighted buttons. In the event that defibrillation is unsuccessful, or if heart rhythm changes, you should stop and request the CMO, who will consider alternative intervention.

3   ADVANCED LIFE SUPPORT PROCEDURES

3.1   Telemedicine

The mass and volume constraints imposed by spacecraft mandate the use of minimal medical equipment, which is compensated for to a degree by the use of telemedicine, a system that uses communications and information technologies for the delivery of clinical care [4]. The network technology downlinking video and audio information makes telemedicine a tool that doctors can use to diagnose health problems in space and even assist the CMO during an operation using space-adapted portable medical equipment.

To date, no operative procedures have been required in space, though operative procedures have been performed with degrees of success in the microgravity environment of parabolic flight [1, 6]. The ability to perform these procedures has, in part, depended on successfully restraining the patient (often using ingenious systems), the surgical equipment, and those performing the operation [2]. The medical training that you and the CMO receive for orbital flight provides only a cursory introduction to the demands of medical care onboard a spacecraft in orbit. Since there will be no physician onboard in the near future it will be the CMO's responsibility to deal with any medical contingencies, but even the CMO's training will permit him/her to perform only minor operative procedures. In the event of a serious operation being required, the operator will need to decide whether to conduct a medical evacuation of the patient, or to rely on the skills of the ground-based flight surgeon in guiding the CMO through an operation using telemedicine. Since a medical evacuation is prohibitively expensive it is likely that operative procedures will be conducted using this remote technology, a situation that will require assistance from one or more spaceflight participants!

In the event of such a situation you may be required to assist in the administration of local and intravenous anesthetic agents, such as barbiturates, benzodiazepines, and narcotics. You may also be required to assist in intubation and endoscopic procedures and various other operative interventions under the direction of the CMO. To prepare you for such an eventuality you will receive training in the basic telemedicine techniques that will include virtual reality *in situ* treatment simulation, vital data acquisition using onboard communications, interactive devices, crew resource management, and advisory and automated diagnostics. After acquiring these basic skills you will have the opportunity to apply them using the HPS in a simulated inflight emergency.

3.2   Ultrasound

One of the techniques you will become familiar with is ultrasound/trauma sonography (TS), a medical tool that provides significant additional diagnostic information and represents a first-choice modality for a large number of higher probability medical scenarios.

One example of the application of this technique is the diagnosis of a problematic medical condition such as an intra-cavitary hemorrhage or determining visceral leakage. Since TS is a noninvasive, fast, safe, effective, repeatable, and tele-transmitting imaging tool, it is possible for such determinations to be made relatively easily. During training you will have the opportunity to use ultrasound to perform basic diagnostic techniques and also learn effective management of ultrasound-supported medical scenarios.

4 MEDICAL CRITERIA FOR EVACUATION FROM ORBIT

Crew return is a last-ditch option reserved for critical failures onboard, lost ability to supply the crew, or medical emergency, of which the latter is probably deemed the most likely. Your operator will determine the parameters, which will mandate a return to Earth, but it is likely that the following criteria will be applied:

1. A disease that has not been cured in flight with available medical pharmacological intervention and which threatens the health of other crewmembers.
2. An epidemic prognosis or clinical symptoms of a highly contagious infectious disease that cannot be treated with available pharmacological intervention.
3. Contamination of the environmental control and life support system by virulent or toxic bacteria that cannot be removed in the presence of humans.
4. Psychological problems with the potential to threaten the health of the sick crewmember or other crewmembers.
5. A life-threatening disease with potential to render or disable a crewmember.

Technical criteria for evacuation

1. Prediction of radiation exposure beyond maximal admissible limits established for a specified mission duration.
2. Unrecoverable drop in partial pressure of oxygen to 140 mmHg.
3. Unrecoverable rise in partial pressure of oxygen to 350 mmHg.
4. Unrecoverable rise in carbon dioxide partial pressure to 20 mmHg that persists for more than 2 hours.
5. An increase in cabin temperature above 33°C that persists for more than 3 days.
6. A decrease in cabin temperature below 15°C that persists for more than 3 days.
7. Any condition in which toxic compounds exceed admissible levels.

5 INFLIGHT HEALTH EVALUATION

During your flight you will have regular access to the flight surgeon who will conduct a health evaluation every three days. During the health evaluation you will have the opportunity to conduct a private videoconference (videocon) with the flight surgeon to discuss any problems you may be experiencing adapting to microgravity. The videocon will be cryptographically protected using encoding and cryptorouters (which will be installed in your issued notebook), which will ensure a confidential information exchange. The health evaluation will require you to be physically examined by the CMO to determine anthropometric measurements and to perform a 12-lead ECG and basic blood biochemistry. This information will then be downlinked to the flight surgeon during communication passes for evaluation.

If you have opted for an EVA experience you will be required to undergo a medical and submaximal fitness check prior to conducting and following completion of the EVA. The pre-EVA check will be conducted the day prior to you performing the EVA and will evaluate your physical performance by means of a basic fitness test. It will also assess the strength of your arm muscles using a grip dynamometer and also a handcrank ergometer. During your EVA you will be instrumented for ECG, BP, HR, and your urine biochemistry will be analyzed in addition to your EVA suit's functional parameters. Following ingress into the vehicle the CMO will perform a routine inspection of your extremities for signs of frostnip and perform urine biochemistry testing.

During your flight you will be encouraged to submit your personal medical information which will be collected, processed, and stored in your notebook. The files will be stored in a memory unit and downlinked automatically to the terminals of the medical operations team when the spacecraft is within coverage of ground tracking stations. Your medical data will add important information to the growing database of how the body reacts to the microgravity environment. For example, during the flight you will be required to monitor your radiation exposure using a radiation dosimeter. The data obtained from this instrument will be used to develop a method of forecasting influences of charged particles and dose rate from space radiation in your particular spacecraft and will be used as a comparison for future spaceflight participants. The subject of radiation protection is covered in Section 6 (below). You will be taught how to use the dosimeter during your lectures on the subject of radiation. Other research that you may have the option of performing may include the resting bioelectric activity of the heart, an investigation that will simply require you to wear an Actiwatch for one hour a day and then to downlink the data on your notebook.

6 RADIATION PROTECTION

Protection limits

The Space Science Board's Committee on Space Medicine recommended the first radiation guidelines for U.S. manned spaceflight in 1970. The panel considered it reasonable to recommend limits based on doubling the natural risk of cancer over a period of 20 years, which resulted in a 2.3% lifetime excess risk. This resulted in an astronaut career limit of 4 sV with an annual limit of 750 mSv being recommended. By way of comparison, your Earth-bound annual exposure to all natural radiation is equal to 3 mSv [7], or 250 times less than an astronaut's annual limit! As new radiobiological data were analyzed it became necessary to reappraise the guidelines and in 2000 the National Council on Radiation Protection and Measurements (NCRP) Scientific Committee produced NCRP Report 132 [3], which determined that a lifetime excess risk of fatal cancer due to radiation exposure of 3% was reasonable. The NCRP findings will likely determine the radiation exposure guidelines your operator will employ and will require that certain measures, such as the ones discussed here, be implemented.

Shielding

Radiation exposure during space missions can be reduced by appropriate measures, but it cannot be prevented as space radiation deeply penetrates all types of shielding and more shielding cannot be provided due to mass constraints.

Among the non-natural means of protecting astronauts from radiation, the simplest solution is to minimize the time spent in space, since radiation dose is a cumulative effect and a long duration in space will obviously result in a higher total dose.

The most common way of providing extra protection against radiation is shielding. However, even though it is usually beneficial, shielding is not always good. Most types of shielding easily stop low-energy particles, but high-energy protons and neutrons as well as other high-energy particles (HZE) are able to penetrate deep into most types of material and during the passage they produce a swarm of secondary particles that contribute to the total radiation dose.

Other possible methods, but somewhat more difficult to use in practice, are screening, prevention, and intervention. By screening potential astronauts for their genetic predisposition to cancer, individuals with low risk of developing cancer can be selected for missions where high doses of radiation are expected. Prevention involves the administration of pharmaceuticals as radioprotectants. Unfortunately, known substances for radioprotection have serious side effects and are thus not suitable for real missions or for spaceflight participants who have paid U.S.$5 million for their flight!

The materials that provide the most effective shielding are those that contain a lot of hydrogen such as polyethylene. On Space Shuttle missions STS-81 and STS-89, data were collected to compare the shielding properties of aluminum and polyethylene, an investigation that found that polyethylene was 30% more effective than a comparable thickness of aluminum as an absorber of radiation from HZE particles. Beyond that point, however, the law of diminishing returns sets in as increasing the thickness of shielding by another three inches provides only about half as much added protection as the first three inches.

Radioprotective and chemoprotective agents and diet

Nausea and vomiting are both common symptoms following exposure to moderate to high doses of radiation, and the use of antiemetic agents is a clinically accepted practice in radiation medicine [7]. Although the development of these symptoms is unlikely during a low Earth orbit (LEO) mission, a solar storm may result in crewmembers suffering significant adverse health effects. In such a situation, given the potential risks associated with exposure to radiation during solar storms, your operator may consider the use of radioprotective agents warranted.

Possible agents that your operator may consider using in the event of exposure to radiation is a combination of amifostine (WR-2721) and its active metabolite, aminopropyl (WR-1065), which can be administered up to 3 hours following exposure. The synergistic effect of these two compounds is to reduce the mutagenic and carcinogenic effects of radiation, although it is not known whether it would be effective in preventing effects produced by protons and heavy ions. Unfortunately, daily use of this compound is not possible since multiple low doses eventually result in toxic effects.

Another possibility currently being investigated is the use of antioxidant nutrients. Results from animal studies indicate that antioxidant nutrients such as Vitamin E and selenium compounds, are protective against lethality and other radiation effects but to a lesser degree than most synthetic protectors. Although some antioxidant nutrients and phyto-chemicals have the advantage of low toxicity, to date they have only been studied in animal models that have employed gamma, X-ray, and high-intensity sources of radiation.

Solar Particle Event Warning System

A part of the CHCS is the Solar Particle Event Warning System (SPEWS), which provides information about the state of flare occurrence and its consequent bursts of electromagnetic radiation, which can reach Earth in about 8 minutes. SPEWS also provides information concerning the particles generated by flare-generated solar storms, which take 18 minutes to arrive.

Although SPEWS is a useful instrument to have onboard, its ability to predict the occurrence of SPEs due to flares is imprecise, and the accuracy of the prediction varies inversely with the length of the advance warning required. Also, the probability of particles resulting from a solar storm reaching Earth depends on the location of the flare on the Sun's surface.

6.1   Radiation exposure

As spaceflight participants, you will be classified as radiation workers and therefore your operator will be required to comply with Occupational Safety and Health Administration (OSHA) regulations related to ionizing radiation exposure. Although no OSHA standards exist for spaceflight, supplementary standards are followed in accordance with 29 CFR 1960.18 [8] which has the following requirements:

a. Its use applies to a limited population.
b. Maintenance of detailed flight crew exposure records.
c. Preflight hazard assessment.
d. Planned exposures to be kept as low as reasonably achievable (ALARA).

e. Maintenance of operational procedures to minimize chance of excessive exposure.
f. Human-made onboard radiation exposure complies with 29 CFR 1910.96 [8].

Monitoring crewmember radiation is a key requirement of your operator's operations, which requires the following three activities:

a. *Preflight*. Activities include projecting mission doses and reviewing crew health records. Prior to flight you will submit a blood sample for analysis. The blood sample will be divided into four parts, which will then be exposed to four different dose levels of gamma radiation. The blood will then be processed and photographs taken of the chromosomes from the cells.
b. *Inflight*. During each mission, continuous radiological support and space environment monitoring will be provided by your operator's mission control.
c. *Postflight*. Crew dosimetry is retrieved and analyzed. On return you will submit another blood sample and chromosome damage counts will be taken; from this information it will be possible to determine the equivalent dose due to radiation received while in space.

Your operator's Radiation Health Protection Program will be administered by your operator's Medical Sciences Division, the responsibilities of which are summarized below:

a. Supports the flight surgeon in advising the Flight Director (FD) during radiation contingencies such as SPE.
b. Maintains astronaut health records including documentation of both mission and medical radiation exposure histories.
c. Provides preflight mission health risk analysis.
d. Establishes radiation health requirements for manned spaceflight based on reviews of current research.
e. Conducts and administers fundamental research into the biological effects of space radiation.
f. Provides operational dosimetry support for crewmembers.
g. Improves and develops engineering tools for use in space radiation exposure analysis.
h. Develops advanced radiation-monitoring equipment.

6.2 Radiation monitoring

To ensure you are not exposed to excessive radiation, measurements are required. On Earth, in environments where high levels of radiation are expected, personnel are equipped with dosimeters which measure the integrated dose of radiation, the value of which can then be compared with given guidelines. Onboard the spacecraft you will also use dosimeters to monitor the cumulated radiation dose, although the dosimeters will be different from those on Earth, as using a general dosimeter is

Figure 6.7. Matroshka. Image courtesy: NASA.

practically impossible due to the range of energy levels of the particles as well as the different types of particles. In order to determine the actual accumulated dose of radiation in space, several types of dosimeters are required to allow a greater range of energies and particle types to be analyzed. Onboard the ISS, in an attempt to quantify this amount of radiation, NASA, the Russian Space Agency, and ESA use a phantom that resembles a human in simple geometric form. Named Matroshka (Figures 6.7 and 6.8), the phantom has several radiation detectors placed inside and outside that allow researchers to determine dynamic radiation doses. Although it is possible that some of the data derived from Matroshka may be used to determine radiation dosages for you and your crewmembers, the radiation-monitoring equipment used onboard your vehicle/habitat to quantify the crew effective dose equivalent will probably not be in the shape of a phantom. Instead, you will probably use one or more of the following items of equipment:

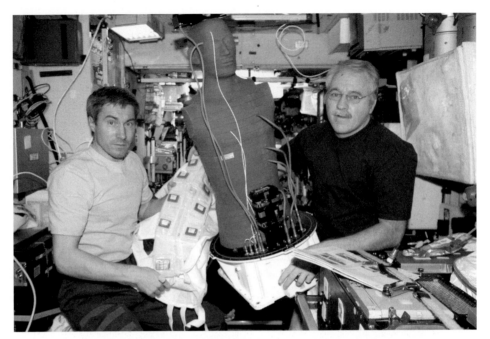

Figure 6.8. ISS crewmembers Sergei Krikalev and John Philips with Matroshka. Image courtesy: NASA.

a. Tissue Equivalent Proportional Counter (TEPC). This is a gas proportional counter that measures the LET spectrum of the incident radiation in a simulated small volume of tissue. TEPC uses a cylindrical cell filled with low-pressure propane gas, hydrocarbon gas that is used to simulate the hydrocarbon content of a human cell. A plastic jacket covering the cell simulates the properties of adjacent tissue cells. Particles passing through the gas release electrons, which are collected by a positively charged biased wire as a pulse. The characteristics of the pulse help identify the energy of the incident particle.

b. Charged Particle Directional Spectrometer (CPDS). This unit measures the flux of all trapped, galactic cosmic radiation and secondary radiation as a function of time, charge, energy, and direction. This device is used to investigate the effect of SEUs on the spacecraft's computers; when charged particles that interact with semiconductor devices generate sufficient charge to change the state of circuits, computer memory is affected and "soft" upsets (requiring reloading of data) or "hard" upsets (resulting in physical damage to the circuit) may occur.

c. Plastic Nuclear Track Detector (PNTD). These are thin sheets of plastic similar to the material used in eyeglass lenses. The PNTD surface becomes pitted with tiny craters as heavily charged ions pass through it. The detectors are returned to Earth after each mission and the plastic is etched to enlarge the craters, which are then counted and their shapes and sizes analyzed using a microscope.

This information is then used to improve the accuracy of the radiation dose the passive dosimetry units have recorded and to improve the estimate of the biological effects of the radiation.

d. Passive dosimetry. The primary unit onboard your spacecraft will probably be the Thermoluminescent Detector (TLD), a flexible, easy-to-use radiation-monitoring system that complements the TEPC and CPDS units. Each TLD resembles a fat fountain pen, which contains calcium sulfate crystals inside an evacuated glass bulb. The crystals absorb energy from incident ionizing radiation as the radiation passes through them. This process results in a steady increase in the energy level of the electrons in the crystal.

References

[1] Campbell, M.R.; and Billica, R.D. A review of microgravity surgical investigations. *Aviation, Space and Environmental Medicine*, **63**, 524–528 (1992).

[2] Campbell, M.R.; Billica, R.D.; Jennings, R.; and Johnston, S. Laparoscopic surgery in weightlessness. *Surg. Endosc.*, **10**, 111–117 (1996).

[3] Fry, M.R.J. *Radiation Protection Guidance for Activities in Low-Earth Orbit.* NCRP Report No. 132 (December 2000).

[4] McCuiag, K.E. Surgical problems in space: An overview. *Journal of Clinical Pharmacology*, **34**, 513–517 (1994).

[5] McCuiag, K.E.; and Houtchens, B.A. Management of trauma and emergency in space. *Journal of Trauma*, **33**, 610–625 (1992).

[6] Stazhadze, L.L.; Goncharov, I.B.; Neumyvakin, I.P.; Bogomolov, V.V.,; and Vladimirov, I.V. Anesthesia, surgical aid and resuscitation in manned space missions. *Acta Astronautica*, **8**, 1109–1113 (1981).

[7] *www.bt.cdc.gov/radiation*

[8] *www.osha.gov/pls/oshaweb/owadisp.show_document?p_table = STANDARDS&p_id = 11274*

Module 4   G tolerance and high-altitude indoctrination training

This section provides you with an introduction to the physiological consequences of sustained acceleration and exposure to hyperbaric conditions, and also introduces you to the practical elements required for completion of this module. The physiology and pathophysiology of increased G and the hyperbaric environment are presented as they relate to problems that you, as a spaceflight participant, may experience during launch, orbit, and re-entry and also to the experience of sustained G and high-altitude indoctrination (HAI) training.

Contents

1   RAPID DECOMPRESSION PHYSIOLOGY

1.1   Introduction

Perhaps one of the most terrifying of the several hazards you will face on-orbit is the one posed by a rapid or explosive decompression of the spacecraft. While a pressure leak or a small perforation caused by a micrometeoroid strike will cause a rapid decompression such as the one experienced by the Mir cosmonauts, a substantial perforation of the skin of the spacecraft will result in an explosive decompression that may have grave consequences for all onboard. It is important therefore that each crewmember is familiar with the physical characteristics of rapid/explosive decompression so that they will be prepared to deal with such a potentially fatal on-orbit event.

1.2   Cabin pressures and physiological responses

In the event of a small-scale rapid decompression it will be necessary to ensure that the cabin atmosphere is maintained below 3,000 meters since the air pressure below this altitude provides sufficient oxygen to maintain normal physiological function without the aid of special protective equipment. Any problems associated with such a leak will be minor ones such as trapped gas problems that will cause only mild discomfort.

   If the rapid decompression has caused a significant volume of air to escape you may be exposed to cabin pressures that require the donning of oxygen masks. If the barometric pressure inside the spacecraft has fallen to between 3,000 m and 15,000 m it will cause noticeable physiological deficits in those crewmembers not wearing oxygen equipment.

   In an explosive decompression event you will need to don pressure suits immediately in order to survive and prevent severe cardiopulmonary damage. Once cabin pressure is reduced to below 19,000 meters (47 mmHg), body fluids begin to boil (ebullism) in the unprotected individual with predictably fatal consequences.

1.3   Respiratory physiology

This section introduces you to the essential laws of respiratory physiology, an understanding of which will enable you to better understand the events associated with your HAI training and with the consequences of rapid and explosive decompression.

Respiratory function

Respiratory function is comprised of five distinct steps. The process of *ventilation* is one in which the pulmonary alveoli exchange gas with the atmosphere. This gas is then exchanged between the alveoli and pulmonary capillaries in a process called *pulmonary diffusion* and transported from the lungs to the tissues and back to the lungs via the vascular system in a process called *transportation*. Gases exchanged between systemic capillaries and tissue cells occur due to *tissue diffusion* and, finally, *cellular utilization* describes the process of chemical reactions that occur within cells that utilize oxygen.

Laws of gas mechanics

It is also useful to understand how the gases we breathe are affected by differences in volume, partial pressure, temperature, and total pressure. To do this we need to review the gas laws, which explain how the properties of volume, temperature, and pressure are implicated in a rapid or explosive decompression event.

The means by which volume affects a gas is explained by *Boyle's Law*, which states that the volume of a given quantity of gas varies inversely with absolute pressure if the temperature remains constant, a relationship that becomes important when we consider the effects of trapped gas.

Partial pressure (PP) is explained by *Henry's Law*, which states that the amount of gas in a solution varies directly with the PP of that gas over the solution. In other words PP is the individual pressure exerted independently by a particular gas within a mixture of gases, as outlined in Table 6.22. The PP exerted by each gas in a mixture equals the total pressure multiplied by the fractional composition of the gas in the mixture, a relationship that is implicated in the various manifestations of decompression sickness (DCS).

The effect of temperature is explained by *Charles' Law*, which states that the pressure of a gas is directly proportional to its absolute temperature, volume remaining constant. The contraction of gas due to temperature change at altitude does not compensate for the expansion due to the corresponding decrease in pressure, a relationship that becomes immediately apparent when experiencing a rapid decompression.

Finally, *Dalton's Law*, which helps us to understand the condition of hypoxia, explains that the total pressure exerted by a mixture of gases is equal to the sum of the partial pressures, which each component would exert if placed separately in the container.

Table 6.22. Partial pressures at different altitudes.

Altitude (m)	Pressure (mmHg)	Oxygen (mmHg)
Sea level	760	160
303	732	132
2,424	564	118
5,454	380	80
10,300	187	39

Oxygen transport

Oxygen is transported in the blood in physical solution and by combining with hemoglobin (oxyhemoglobin). A gram of hemoglobin has an oxygen-carrying capacity of 1.34 ml, which equates to a capacity of 20 ml of oxygen per 100 ml of blood assuming normal hemoglobin content of 14.5 g/100 ml. This represents 100% oxygen saturation.

Hypoxia

Hypoxia is caused by the lower atmospheric pressure at high altitudes (i.e., a reduced arterial oxygen pressure, or PaO_2), which inhibits the diffusion of oxygen from the air to the lungs. Consequently, less oxyhemoglobin is produced, resulting in decreased oxygen transport to the tissues, which results in the signs and symptoms described in Table 6.23.

Perhaps the most dangerous sign of hypoxia is impaired judgment, since, even if symptoms are experienced, a crewmember may disregard them and not take corrective action, which could prove hazardous. Although your spacecraft will use state-of-the-art oxygen delivery systems, be fitted with advanced cabin pressurization technology, and have installed advanced life support equipment, a small puncture in the vehicle's skin will rapidly cause the above symptoms. It is vital therefore that each crewmember is familiar with the subjective and objective effects of hypoxia.

A faulty pressurization in your spacecraft may cause you to experience a gradual onset of signs and symptoms as described in Table 6.23, but in the event of an explosive decompression you will experience symptoms of hypoxia almost immediately. Such an event will require you to take immediate action since in an explosive decompression situation cabin pressure may be reduced to a level that does not support life within seconds (Table 6.24).

1.4   Trapped gas

As you ascend to altitude during your chamber run, the free gas in your body cavities will expand and if the escape of this gas is impeded, pressure will build up within the

Table 6.23. Signs and symptoms of hypoxia.

Signs of hypoxia	Symptoms of hypoxia
Rapid breathing	Air hunger
Cyanosis	Dizziness and headache
Impaired coordination	Mental and muscle fatigue
Lethargy/Lassitude	Nausea
Executing poor judgment	Visual impairment

Table 6.24. Time of useful consciousness (TUC)/Effective performance time (EPT).

Cabin pressure (mmHg)	Equivalent altitude (m)	Ambient PO_2 (mmHg)	Conscious time
349.5	6,060	73.22	5–12 min
282.5	7,575	59.2	2–3 min
216.1	9,090	45.4	45–75 s
179.3	10,605	37.6	30–60 s
141.2	12,120	29.6	10–30 s
111.1	13,635	23.3	12–15 s
87.5	15,150	18.3	<12 s
26.6	22,725	5.6	<12 s
8.36	30,300	1.8	<12 s
1.48	60,600	0.3	<12 s
9.49^{-4}	90,900	<0.1	<12 s
6.31^{-6}	136,350	<0.1	<12 s

TUC = time between an individual's deprivation of oxygen at a given altitude and the onset of physical and mental impairment that compromise his/her taking rational action.

cavities and you will experience pain due to the difference between internal body gas pressure and external chamber pressure. This process is described by Boyle's Law and will cause gases trapped in your body cavities to expand as altitude increases, and to contract as altitude decreases. The cavities that will primarily be affected during your chamber run will be your ears, sinuses, and gastrointestinal system. This section examines some of the effects you may experience.

Ear block

Ear block, the symptoms of which include pressure in the ear, dizziness, or ringing in the ear, may occur during ascent or descent when the air pressure in the middle ear is unable to equalize with the ambient pressure. This may be caused a cold/sinus infection or an occlusion of the Eustachian tube due to ineffective equalizing procedures, such as pinching the nose shut while swallowing with the chin on the chest. Ear block may also be caused by too rapid an increase in barometric pressure when returning from altitude or from having breathed 100% oxygen.

If you suffer from this condition, try swallowing, yawning, or tensing the muscles of your throat, a procedure that will open the orifices to the Eustachian tube, thereby allowing equalization. You can also try performing periodic Valsalva maneuvers during the first one to two hours post-flight, which will lower the oxygen concentration by flushing the middle ear with ambient air.

Sinus block

The sinuses are air-filled, bony cavities lined with mucous membranes that connect the nose, and if these become obstructed, either as a result of infection or swelling, you may experience an almost incapacitating pressure and/or pain in the frontal sinus. If you are unlucky enough to suffer this condition you should perform frequent Valsalva maneuvers during the descent and request the chamber run is terminated if pain is experienced during ascent. This condition is often precipitated by a cold infection, so you should not participate in a chamber run to altitude if you have an infection.

Barodontalgia

Barodontalgia is altitude-related toothache, a condition invariably associated with pre-existing dental pathology, often entrapped air under imperfect fillings or pre-existing dental conditions such as an abscess. Symptoms of this condition may include barosinusitis or pain in the affected tooth, although high pressure under a filling may cause excruciating pain, and in rare instances, cause the tooth to explode! The condition can be prevented by not undertaking excursions to altitude when suffering from pulpitis or carious teeth and reducing the rate of increase of barometric pressure if symptoms are noticed.

Trapped gas disorders of the gastrointestinal (GI) tract

This condition is caused by the digestive processes of fermentation and decomposition of food undergoing digestion. GI problems associated with rapid decompression range from merely embarrassing to totally incapacitating. Typical symptoms include GI pain, GI sensitivity and/or irritability, and in extreme cases loss of consciousness. Fortunately, avoiding this condition is relatively simple and requires you, prior to your excursion to altitude, to avoid gas-producing foods such as beans and cabbage, ensure you have thoroughly masticated your food, and avoid drinking large quantities of liquid.

1.5   Altitude decompression sickness

Henry's Law states that when the pressure of a gas over a liquid is decreased, the amount of gas dissolved in that liquid will also decrease. In the body, nitrogen is an inert gas stored in physical solution, and in an event such as an explosive decompression your body will be exposed to decreased barometric pressure which will cause dissolved nitrogen to be forced out of solution. If the nitrogen is forced to leave too rapidly, bubbles will form in areas of your body, causing a variety of signs and symptoms, the most common of which is joint pain, also known as altitude decom-

Table 6.25. Signs and symptoms of altitude decompression sickness.

DCS type	Bubble location	Signs and symptoms
BENDS	Large joints of the body (shoulders, elbows, knees)	Deep localized pain. Occasionally a dull ache. Active and passive joint motion aggravates pain. Pain may occur at altitude, during descent, or several hours later
NEUROLOGIC	Brain	a. Confusion and/or memory loss b. Scotoma (spots in visual field), double vision (diplopia), or blurry vision c. Seizures, fatigue, dizziness, vomiting, vertigo
	Spinal cord	a. Burning, stinging, and tingling around lower chest and back b. Symptoms may spread from feet up and be accompanied by ascending paralysis/weakness c. Abdominal or chest pain
	Peripheral nerves	a. Numbness, stinging, paresthesias b. Muscle weakness
CHOKES	Lungs	a. Deep chest pain aggravated by breathing b. Dry constant unproductive cough c. Shortness of breath (dyspnea)
SKIN BENDS	Skin	a. Itching around face, neck, and arms b. Crawling sensation over skin c. Mottled/marbled skin around shoulders/chest

pression sickness (DCS), or "the bends". DCS and its symptoms may, in severe cases, result in shock, and if treatment is not promptly administered, death.

Typical signs and symptoms of DCS are listed in Table 6.25. Most bubble formation sites relate to joint pain, but between 10% and 15% of cases will also present neurological symptoms such as visual disturbances. For those suffering from altitude DCS the immediate treatment is to bring the victim down from altitude for medical evaluation and administration of hyperbaric oxygen therapy consisting of 100% oxygen.

DCS can be prevented by conducting an oxygen prebreathe, which requires breathing 100% oxygen for 30 minutes in order to purge nitrogen (termed "washout") from the body.

There are a number of factors that predispose an individual to DCS, one of which is the rate of ascent, the faster the rate usually increasing the risk. Time spent at altitude is also a factor, since the longer the duration of exposure to altitudes greater than 5,500 meters, the greater the risk of altitude DCS, although DCS is rare in exposures of less than five minutes. A surface interval of less than 3 hours between exposures to altitudes over 5,500 meters may also increase DCS risk. Exposure to hyperbaric conditions such as scuba diving increases the rate of off-gassing, which is why individuals are not exposed to hypobaric conditions within 24 hours following diving. Age is also a predisposing factor since the incidence of altitude DCS in individuals aged 40 to 45 years is three times that of those aged 19 to 25. Because exercise results in an increase in muscle perfusion and inert gas uptake (both mechanisms that increase susceptibility to DCS) individuals being exposed to altitude should refrain from strenuous exercise for 12 hours prior to exposure and for 6 hours following exposure. There is also a gender component, the incidence of altitude DCS in females being almost twice that of men.

1.6   Rapid/Explosive decompression

The difference between a rapid and an explosive decompression is simply related to time. A decompression that occurs in less than half a second is an explosive decompression, whereas a rapid decompression is one that takes more than half a second, but less than ten seconds. In the latter type there is less potential for fatal physiological damage, although the physical characteristics will still be present.

Physical characteristics of rapid/explosive decompression

In order to determine if a decompression event has occurred, it is necessary to be familiar with certain physical and observable characteristics. The first sign that the spacecraft has been penetrated is noise. If the size of the penetration is relatively small, air that is escaping into a vacuum will make a "swish" sound, whereas a large penetration will normally result in a loud explosive sound. The explosive sound will be preceded by a loud popping sound, like the sound of champagne cork, only a hundred times louder.

a.   Fogging. You will experience this phenomenon during the rapid decompression to 25,000 meters. It is explained by Charles' Law and is caused by the sudden change in temperature and/or pressure changing the amount of water vapor the air is able to hold. In a rapid decompression, temperature and pressure are reduced, which reduces the holding capacity of air for water vapor, so water vapor that cannot be held by the air appears as fog.

b.   Temperature. In the event of decompression, cabin temperature will equalize with the external ambient temperature (close to absolute zero!). This will result in a significant risk of frostbite and cold-related injuries.

c.   Flying debris. The magnitude of the decompression force will depend on the size of the puncture, but given the high pressure differential it is likely that the velocity of airflow through the opening will force unsecured items to be extracted, and in the event of a large opening it is possible crewmembers may be sucked from the spacecraft! For those who would like to see a graphic example of such an event, the Danny Boyle film *Sunshine* is highly recommended!

Physiological effects of rapid/explosive decompression

In the event of an on-orbit decompression, crewmembers will be faced with several life-threatening effects on the pulmonary, cardiovascular, gastrointestinal, and CNS, most of which are summarized below:

a. Pulmonary system. This system is potentially the most vulnerable during a rapid decompression, due to the large volume of air in the lungs and the fragile nature of pulmonary tissue. The extent of the damage inflicted on the system will depend on the magnitude and rate of decompression, which in turn will determine survivability [4, 8]. If the decompression is rapid rather than explosive, crewmembers may experience mild to moderate pulmonary hemorrhaging and edema [5], whereas if the decompression event results in a decompression rate faster than the capability of the lungs to decompress there will be a transient positive increase in intrapulmonary pressure with potentially fatal consequences.

b. Cardiovascular system. Damage inflicted on the heart will primarily be caused by anoxia, which results in stretching of the myocardium [1]. After approximately 30 seconds, heart rate will decrease to almost half of resting levels, although cardiac contractility will be maintained [3, 6]. As cabin pressure approaches vacuum, heart rate will drop significantly.

c. Central nervous system. Most of the effects in a rapid decompression situation will be due to anoxia and its effect on brain function [2]. As less oxygen is available, the CNS will become progressively more damaged and the crewmember will suffer greater neurological damage.

d. Gastrointestinal system. The dangers associated with this system are related to the rapid expansion of trapped gases within the body cavities. The abdominal distension that results from rapid decompression will displace the diaphragm, which will compromise respiratory function to such a degree as to impede breathing. Ultimately, as cabin pressure approaches vacuum, blood pressure will drop precipitously and unconsciousness and shock will ensue.

e. Hypoxia. A rapid reduction in ambient pressure will produce a corresponding drop in the partial pressure of oxygen (Table 6.27) and, depending on the magnitude of the drop, a performance decrement as previously discussed.

f. Hypothermia. The protective clothing worn and the speed at which cabin temperature drops will determine the severity of cold-related injuries such as hypothermia and frostbite.

1.7 Medical interventions following rapid decompression

There is no treatment protocol for exposure to rapid or explosive decompression, although viable options are available. A crewmember that has suffered a rapid/explosive decompression will be brought to the airlock which will serve as a hyperbaric chamber for recompression. Recompression may be successful in reversing the massive tissue swelling associated with an explosive decompression and may permit further treatment of the victim. In this scenario, the crewmember will be placed inside the airlock and hyperbaric oxygen therapy will be administered by the CMO in accordance with established protocols. If pulmonary hemorrhage is evident and respiratory function is compromised, the CMO may request assistance in order to administer endotracheal suctioning and intubation [7]. If the crewmember is diagnosed with internal bleeding, the CMO will insert two large-bore IVs in order to administrate Dextran, a fluid expander designed to offset plasma loss. In the event of a seriously injured crewmember, the CMO may need to administer advanced cardiac life support.

2   HIGH-ALTITUDE INDOCTRINATION

The HAI training phase will comprise a theoretical component delivered at your operator's facility followed by a practical component at the NASTAR facility. During your initial exposure to high altitude you will breathe supplemental oxygen for 30 minutes at ground-level pressure to reduce the nitrogen load in your body tissues, a procedure termed "prebreathing". To provide you with an opportunity to experience hypoxia, the chamber will be gradually depressurized to an altitude of 7,500 meters. You will then remove your oxygen masks and complete simple, repetitive drawing and mathematical tasks for a period of up to four minutes, after which you will be repressurized and returned to ground pressure.

2.1   Ground school

The ground school was mostly covered in the theoretical modules described in the first part of this module, although on arrival at NASTAR your instructors will provide you with a quick review.

Following the review a hyperbaric technician will give you a tour of the hyperbaric chamber, after which you will have the opportunity to experience a dry chamber run. Following the dry run you will be assigned a seat and an instructor will provide an overview of what to expect at each flight level while the chamber technicians will supervise you connecting and disconnecting your built-in breathing system (BIBS).

2.2   Practical preparation

Two chamber operators will accompany you and the other three spaceflight participants when you enter the chamber. The chamber operator will be responsible for communicating with the two console operators seated outside and also for controlling the interior pressure from start to finish. They will also be responsible for sealing the lock leading to the access chamber/entry lock, and ensuring safety procedures are adhered to. During the ascent to altitude the entire chamber will be maintained at the same internal pressure. In the event of someone not feeling well, the affected person will be placed in the entry lock and this will then be sealed from the main chamber. The entry lock will then be depressurized.

During the chamber run the console chamber operators will record the times of the start and finish time of prebreathe, the time each altitude is reached, and the length of time the chamber is kept at each altitude.

2.3 Prebreathe

Thirty minutes prior to the chamber run, you will don masks connected to 100% oxygen and commence your prebreathe. The chamber technicians will then ensure your masks are fitted correctly and conduct a communication check. Following prebreathe the chamber operator will close the chamber's external hatch and the run will commence at a nominal ascent rate of 1,500 meters per minute. At 1,500 meters the ascent will be stopped to check for system leaks. During the ascent the chamber operators will remind you of the physiological effects you would be experiencing if you were not breathing via BIBS as described in Table 6.26.

At 7,500 meters you will buddy up and perform a hypoxia awareness test, which will require you to drop your mask and shut off oxygen panels in order to experience oxygen deprivation. The hypoxia familiarization is an essential element of HAI, as it teaches you to recognize your own physiological response to hypoxia, and also makes you aware of your reaction times. During your exposure to 7,500 meters, you will complete a form similar to that in Figure 6.9 at 60-second intervals. After a maximum time of 4 minutes at 7,500 meters you will be instructed to don oxygen masks and restart your oxygen supply and you will descend to sea level.

Table 6.26. Characteristics of hypoxia.

	Stages of hypoxia			
	Indifferent (98%–90% O_2 saturation)	*Compensatory* (89%–80% O_2 saturation)	*Disturbance* (79%–70% O_2 saturation)	*Critical* (69%–60% O_2 saturation)
Altitude (000s m)	0–3	3–4.5	4.5–6	6–7.5
Symptoms	Decrease in vision	Drowsiness. Poor judgment. Impaired coordination and efficiency.	Impaired motor control and handwriting. Decreased coordination. Impaired vision, intellectual function, and memory	Circulatory failure. CNS failure. Convulsions. Cardiovascular collapse. Death

List your symptoms						Numerical Problems

	Magnitude			
Symptom	**1**	**2**	**3**	**4**
Headache				
Dizziness				
Anxiety				
Difficulty concentrating				
Tunnel vision				
Colors reduced				
Blurred vision				
Tiredness				
Tingling				

Numerical Problems

1. 1313   6792   3695
   -1212   +4981   x5

2. Subtract 7 from 101: 94/87/80/ / / / /
   / / / / / /

3. Write you name:

Eye-hand coordination.

1. Draw a cat.

2. Draw a car.

3. Draw a house.

Response for Minute 1

Simple Questions:
1. Who was the first man to walk on the moon?

2. Who is the Prime Minister of Great Britain?

3. Who was the last American to win the Tour de France?

Figure 6.9. Hypoxia demonstration sheet.

2.4   Exposure to high altitude and rapid decompression

Following your chamber run you will have the opportunity to familiarize yourself with the pressure suit and helmet you will be wearing during launch and re-entry, after which there will be a break for lunch. The suit technicians will begin by asking you to perform simple tasks such as pulling pens from pockets and testing your manual inflation systems. You will then perform a press-to-test, which will allow you to regulate pressure inside the suit. To ensure you are comfortable with the suit you will experiment with dialing in, regulating, and maintaining various pressures, after which the suit technician will perform a final check of your suit and the chamber will be prepared for another run.

2.5   Flight to 25,000 meters

During the ascent you will be reminded about suit inflation and the various events that occur at specific altitudes. For example, as the chamber passes through the *Armstrong Line* at 19,100 metres, you will notice the suede patches on your suit begin to smoke and the water in the glass beaker begin to boil, a not so subtle reminder that if you were not wearing a pressure suit your blood would be boiling! As the chamber approaches 25,000 meters, the instructor will explain the emergency egress checklist and ask you to find and touch each item related to conducting an emergency egress while the suit is fully inflated. The chamber will remain at 25,000 meters for four minutes during which time the chamber operator will explain what to expect during the rapid decompression exercise.

The chamber will descend to 7,500 meters, but the entrance lock will remain at 25,000 meters. At 7,500 meters the chamber operator will give you a briefing about the events preceding rapid decompression. The console operator and chamber operator will verify that the main chamber and the entrance lock are at the correct altitude and the console operator will then open the transfer valve between the two chambers and signal to the chamber operator that it is safe to conduct rapid decompression. The chamber operator will press the red button located in the ceiling of the main chamber and rapid decompression will occur, the first sign of which will be a loud "bang", followed by a forceful rush of air. You will see condensation instantly form on the interior of the chamber, and within seconds you will be unable to see those seated in front of you due to the fogging of the air inside the chamber. The water you observed boil in the beaker will explode out of its container and you will feel your suit become rigid. This is something you definitely do not want to happen while you are on-orbit!

3   G-TOLERANCE THEORY

3.1   Introduction

During the launch, ascent, orbital, de-orbit, and re-entry flight regimes you will encounter different acceleration stresses and although your spacecraft has been designed to reduce these stresses, in the event of a contingency, especially during re-entry, these forces may be very large, and may impose forces in excess of 15 G.

This section provides you with an introduction to the theoretical and practical aspects of G tolerance and the accelerative forces you will encounter during launch and re-entry. The theoretical training for this module will take place during the first two weeks of your training at your operator's facility and the practical phase will be conducted in the human centrifuge at the NASTAR training facility. However, before you can enjoy the centrifuge it is necessary to understand the basic principles that govern gravitational physiology.

Sustained acceleration $(+G_z)$ is acceleration that lasts for more than 1 second and is a force that can make it impossible for you to breathe. High rates of sustained acceleration can also result in blood pooling to such a degree that it will cause you to convulse and eventually black out. Given the serious consequences of these events it is important you are familiar with the effects so you are able to deal with inflight events such as grayout, blackout, or even unconsciousness.

During your trip to and from orbit you can expect to experience five distinct phases of accelerative stress, each differing in their magnitude and duration.

a. *Launch.* Typically between 3.5 G and 4.5 G.
b. *Orbital.* The centrifugal force of the spacecraft balances the gravitational force, thus producing a microgravity environment of zero gravity!

c.   *Re-entry.* You will begin to notice acceleration stresses at an altitude of 75,000 meters due to the sudden drag and deceleration during re-entry into the denser atmosphere. The magnitude of G forces that you experience will depend on your spacecraft's angle of entry into the atmosphere as determined by your operator. High re-entry angles ($>10°$) produce very large forces ($>25\,G$) whereas shallow angles less than $1°$ usually result in forces of less than $5\,G$.

d.   *Landing.* If your operator's vehicle uses big parachutes and you land on soft terrain you should experience landing forces no greater than $5\,G$.

e.   *Emergency egress.* The forces experienced during an emergency egress will be different in different phases of flight, but you can expect high-magnitude accelerations that exceed $15\,G$, sustained for 1 or 2 seconds.

3.2   Cardiovascular effects of $+G_z$

The most sensitive of the physiological systems to $+G_z$ is the cardiovascular system. Before the centrifuge run you will instrumented with ECG and heart rate–monitoring equipment so you will be able to see for yourself how you react to increasing G. Generally, you can expect your heart rate to correlate with increased $+G_z$ due to the acceleration force effect and the general psychophysiological stress syndrome that is associated with exposure to acceleration. In fact, most people will experience an initial cardiovascular response even before the start of the run due to anticipation of the event.

3.3   Respiratory effects of $+G_z$

The rapid onset runs (RORs) will expose you to five or more G and will help indoctrinate you to the effects on your respiratory system that you will experience during launch. The effect of being launched in the reclined position will compress the chest as accelerative (G) forces increase, a sensation that astronauts describe as having an elephant sitting on their chest!

3.4   Sensory effects of $+G_z$

Many of the CNS effects of $+G_z$ are a direct consequence of the cardiovascular effects, since a regular blood supply is required for the CNS to function, so the ability of your body to tolerate acceleration is related directly to adequate blood flowing to your brain. Because of this relationship, symptoms that relate to insufficient blood flow to the brain are used to determine tolerance to $+G_z$. The normal index of defining G-level tolerance is to use loss of vision (LOV) in an upright-seated position at a specific level of G exposure. The visual symptoms you will experience during your centrifuge run are caused by a reduction of blood flow in the retina of the eye, which in turn is caused by a reduction in driving pressure and higher intraocular pressure. Table 6.27 summarizes the sensory symptoms you may experience during your centrifuge run.

You will be able to measure light loss by watching a 71 cm light bar placed 76 cm

Table 6.27. Categorization of light loss criteria.

Symptom	Description	Onset of symptoms (G)	Criteria
Grayout	Partial LOV. Often occurs as first physiological effect of sustained G loads. Low blood oxygen levels cause peripheral vision to fade. Objects in center of FOV can be seen but seem surrounded by gray haze	3.5	100% peripheral light loss (PLL) combined with 50% central light loss (CLL)
Blackout	Gray haze envelops entire FOV and almost immediately becomes black. You will be conscious but unable to see	Above 5	100% CLL, but sufficient blood reaches brain to permit consciousness and hearing
Gravity-induced loss of consciousness (G-LOC)	Follows quickly after blackout with sustained G load. You will be unconscious but will regain consciousness when G load is released	Above 5	Normally occurs following increase of acceleration after blackout

LOV = loss of vision; FOV = field of vision.

in front of you at eye level. The bar has a 2.5 cm diameter green light at each end and a 2.5 cm diameter red light in the center. When you look directly at the light bar without moving your eyes or head and you cannot see the green lights but can see the red light, 100% PLL has occurred (Figure 6.10).

3.5 Individual tolerance to +G$_z$

Tolerance to +G$_z$ may vary from day to day and is highly individualized as everyone has different physiological responses, but there are some steps you can take to minimize any unpleasantness. You should ensure you have eaten prior to the centrifuge run because if you are hypoglycemic you will impair your heart's ability to compensate at the onset of high G loads, and you will experience grayout and blackout at relatively low sustained G loads. If you are unfit you can expect a significant decrease in your ability to tolerate +G$_z$. If you have an illness you should inform the staff at the NASTAR center as most illnesses will compromise your tolerance to +G$_z$. Be sure to drink adequately prior to your run as dehydration will have an adverse effect on your ability to tolerate G by reducing plasma volume.

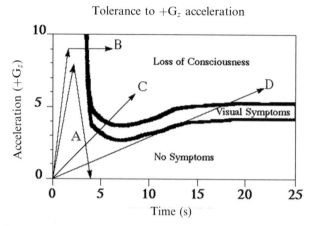

Figure 6.10. Effect of increasing G on vision. Image courtesy: U.S. Navy. Source: *http://nomi. med.navy.mil/NAMI/aeromedicalbriefs/TIP.ppt*

Finally, recency of G exposure will affect your tolerance to $+G_z$, as this declines markedly if frequent exposure to G is not maintained.

4   G-TOLERANCE TRAINING OVERVIEW

4.1   Ground school

On arrival at the NASTAR facility you will review the major theoretical elements of G tolerance and be taken on a tour of the facilities that will include the console room and the centrifuge chamber.

The NASTAR centrifuge has the capability of an onset rate of 6 G per second compared with the 0.1 G/s you will be experiencing. During your indoctrination to the facility you will be shown the interior of the gondola and the instructor will indicate the adjustable rudder pedals provided for foot support, and the shoulder and lap harnesses that will secure you. You will also have the opportunity to don the facemask you will be wearing during the run to monitor your breathing. The face-mask serves a dual function as it also permits two-way communication with the console operator. When you sit in the chair you may notice a small video camera, which will record you during the run. During your gondola indoctrination session you will also be given the opportunity to practice on the peripheral light loss (PLL) system.

4.2   Practical preparation

Shortly after breakfast on test day you will observe a dry run from the console room at NASTAR. The console operator will review the G-onset loads and review the operation of the communication system and will then assign you to a centrifuge

Table 6.28. Run schedule for determination of G sensitivity.

Run #	Type of run	Rate of onset	Peak G		Rate of offset
		(G/s)	Magnitude	Duration (s)	(G/s)
1	Warmup	0.5	2.5	15	0.2
2	GOR	0.1	5.0	5	1.0
3	ROR 1	1.0	3.0	10	1.0
4	ROR 2	1.0	4.0	15	1.0
5	ROR 3	1.0	5.0	20	1.0

rotation and explain the safety procedures, the role of the flight surgeon, and centrifuge operator. Before you step into the gondola the flight surgeon will explain what you should expect during each run. Support personnel will then supervise your ingress into the gondola where you will be connected to biomedical instrumentation that will include a 12-lead ECG, blood pressure cuffs, and respiratory monitoring equipment.

4.3   G-tolerance test

The run schedules for your assessment are detailed in Table 6.28.

References

[1] Burch, B.H.; Kemp, J.P.; Vail, E.G.; Frye, S.A.; and Hitchcock, F.A. Some effects of explosive decompression and subsequent exposure to 30 mmHg upon hearts of dogs. *Journal of Aviation Medicine*, **23**, 159–167 (1952).

[2] Casey, H.W.; Bancroft, R.W.; and Cooke, J.P. Residual Pathological Changes in the Central Nervous System of Dogs Following Rapid decompression to 1 mmHg. *Aerospace Medicine*, **37**, 713–718 (1966).

[3] Dunn, J.E.; Bancroft, R.W.; Haymaker, W.; and Foft, D.W. Experimental animal decompression to less than 2 mmHg. *Aerospace Medicine*, **36**, 725–732 (1965).

[4] Edelmann, A.; Whitehorn, W.V.; Lein, A.; and Hitchcock, F.A. *Pathological Lesions Produced by Explosive Decompression.* WADC-TR-51-191.

[5] Hall, W.M.; and Cory, E.L. Anoxia in Explosive Decompression Injury. *American Journal of Physiology*, **160**, 361–365 (1950).

[6] Kemph, J.A.; and Hitchcock, F.A. Changes in blood and circulation of dogs following explosive decompression to low barometric pressures. *American Journal of Physiology*, **168**, 592 (1952).

[7] Kolesari, G.L.; and Kindwall, E.P. Survival following accidental decompression to an altitude greater than 74,000 feet. *Aviation, Space and Environmental Medicine*, **53**(12), 1211–1214 (1982).

[8] Topliff, E.D.L. Mechanism of lung damage in explosive decompression. *Aviation, Space and Environmental Medicine*, **47**, 517–522 (1976).

Module 5 Space motion sickness and zero-G theory

This section introduces you to the theory of space motion sickness (SMS), teaches you to recognize the signs and symptoms of the syndrome, and provides you with techniques to reduce the debilitating effects of the condition in preparation for your indoctrination to microgravity onboard G-Force-One.

Contents

1 INTRODUCTION TO SPACE MOTION SICKNESS

Two-thirds of first-time spaceflight participants will experience symptoms of SMS, which may include headache, stomach awareness, nausea, and vomiting. You may notice these symptoms shortly after orbital insertion and may find them triggered by viewing an unusual scene such as an inverted crewmember, although symptoms may also be provoked by head movements [1]. The good news is that symptoms normally abate within 48 to 72 hours inflight, although the rate of recovery, degree of adaptation, and specific symptoms vary between individuals.

The development of SMS typically follows an orderly sequence, the timescale largely being determined by factors such as the individual's susceptibility and the intensity of the motion stimulus. Although the incidence and severity will depend on the particular environment involved, SMS is always unpleasant and may in certain situations compromise performance during an emergency egress.

Exposure to provocative motion on Earth generally leads to prodromal signs and symptoms, but this is not always observed in microgravity. SMS, although sharing common features with Earth-bound MS, is a syndrome that has its own set of features and symptom development.

1.1   Space motion sickness symptoms

A common feeling on orbital insertion is one of disorientation, a matter that will probably be addressed onboard your vehicle by careful location of lighting and color schemes to give you a definite "up" and "down" feeling. The sensation of disorientation is likely to precipitate classic SMS symptoms such as GI awareness, perhaps as early as the first few minutes on reaching orbit. These symptoms may range from benign nausea to vomiting and retching. Unfortunately, as already mentioned, these episodes will not be preceded by prodromal symptoms as they are on Earth. Most likely the first indication that you are suffering from SMS will be a forceful expulsion of stomach contents, which you will need to capture and stow as soon as possible to avoid the ire of your fellow crewmembers! If you are among the lucky few that do not experience full-blown SMS it is still likely you will experience milder symptoms such as malaise, lack of initiative, and general irritability. One way for you to reduce the incidence of SMS symptoms is to minimize head movements since hypersensitivity to head motion is perhaps one of the more commonly suffered symptoms and may provoke stronger sensations for reasons that are explained in the following sections.

1.2   Essential neurovestibular physiology

In this section we discuss the main theories of the SMS syndrome and also the basic physiology of the two components that relate to self-motion: the vestibular system and visual perception.

The vestibular system (Figure 6.11) consists of the inner ear, in which are located three semicircular canals for detecting angular acceleration and the saccule and utricle which detect linear acceleration. The semicircular canals correspond to the three dimensions in which movement occurs, each canal being responsible for detecting motion in a single plane. Flowing through each canal is a fluid (endolymph), which deflects small hair-like cells (cupula) as the head experiences angular acceleration, which in turn sends messages to the vestibular receiving areas of the brain. One vestibular component is located on each side of the head, their function being to mirror each other and act in a push (excited)–pull (inhibited) manner depending on the direction the cupula move.

When you move forward (when accelerating in a car, for example (linear acceleration)), this information is communicated to the brain via the utricle and saccule,

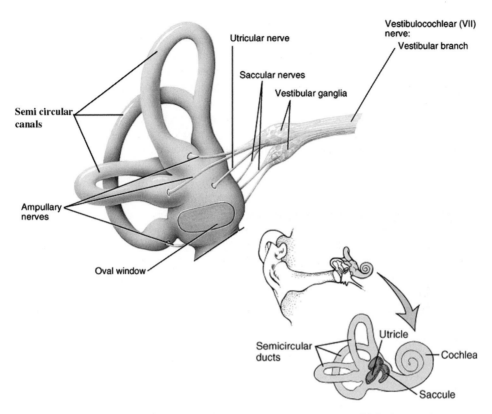

Figure 6.11. Components of the vestibular system. Image courtesy: NASA. Source: *http://web oflife.nasa.gov/learningResources/vestibularbrief.htm*

structures that have a sheet of hair-like cells (macula) embedded in a gelatinous mass. This mass has areas of small crystals (otolith), which provide the inertia required to drag the "hairs" from side to side, thereby providing the perception of motion. Once you have decided on a speed to drive, a steady velocity is detected, the otoliths stabilize and the perceived motion dissipates. The arrangement of the utricle and saccule determine motion detection, the utricle being responsible for motion in the horizontal plane since it lies horizontally in the ear and the saccule able to detect down, up, forward, and backward motion by virtue of its vertical orientation.

Although a familiarity of the vestibular system is helpful if we want to understand SMS and the theories of the syndrome, it is also necessary to appreciate the relationship between this system and the visual system.

1.3   Etiology

One of the problems faced by space medicine experts is that—despite the high incidence of SMS—there is no reliable ground-based test to predict which crewmembers will be affected. Also, despite nearly 50 years of manned spaceflight, little knowledge exists of how to prevent SMS and there are no fully acceptable means of treating SMS symptoms once they appear. Although the underlying causes of SMS syndrome are reasonably well understood, the mechanisms are not clearly defined and no satisfactory methods have been identified for its prediction, prevention, or treatment.

Unsurprisingly, considerable research has been directed at elucidating consistent and objective physical signs that correlate with the onset of the constellation of symptoms associated with SMS, but much of this research has been largely unsuccessful. A part of the reason for the lack of success in accurately identifying methods for its prediction, prevention, and treatment is due to the inflight use of anti-MS drugs, a procedure that tends to obscure meaningful clues. Also, involving large numbers of astronauts are rarely feasible because they are extremely costly and may take considerable time to complete. Though you may have the impression that astronauts are being constantly launched into low Earth orbit, the number of subjects available to be tested is very small in relation to scientific requirements, because of the high variability in the type of flights. Although astronauts serve as an elite subject group, their numbers are small when dealing with the complex nature of SMS. This situation is further complicated by the limited opportunities available to obtain data during spaceflight, and by the number of astronauts participating in a given study being usually only one or two.

2 SPACE MOTION SICKNESS RESEARCH

Due to the limitations of conducting spaceborne research it would seem logical to conduct SMS-related research in terrestrial laboratories. Unfortunately, methods of simulating the syndrome are difficult and have not been fully studied, while the views on the mechanics of its development are quite diverse, and at times, contradictory. One of the most realistic ways to simulate microgravity is by means of a parabolic flight profile, although this method is limited by time considerations as it provides only short periods of microgravity, an explanation of which is provided in this module.

2.1 Vection

As you may be incapacitated for up to three days on orbit, it is important that you undergo training to assist you becoming familiar in recognizing SMS symptoms and also to help you desensitize yourself against these symptoms, thereby reducing the chances of you becoming sick. Unfortunately, much of this training involves you being exposed to motion-provoking environments!

One of the classic methods of inducing experimental MS is to expose seated individuals to whole-field visual stimulation [9] (Figure 6.12). For this test you will be seated in a large drum that will be rotated to present a moving array of vertical black and white stripes [4, 7, 8, 9]. Within 5 to 30 seconds of drum rotation you will probably experience compelling illusory self-rotation, which is termed "vection".

After a few minutes of this rotation it is likely that you will begin to experience symptoms of MS that may become progressively more severe. The reason you will experience self-rotation is due to a hypothesized [11] sensory mismatch or sensory conflict between visual inputs, which indicate self-motion, and vestibular and proprioceptive inputs [5, 6], which indicate no motion occurrence. The sensory conflict you will experience while in the drum is assessed for its conformity with certain patterns established on the basis of your previous experience of motion environments. If the novel sensory input fails to conform to the established patterns of your previous motion experience, the sensory conflict will provoke symptoms of MS.

Following the theory component of this module you will have the opportunity to experience vection shortly following lunch (!). During your exposure you will be seated in a vection drum similar to the one in Figure 6.12, which will rotate at

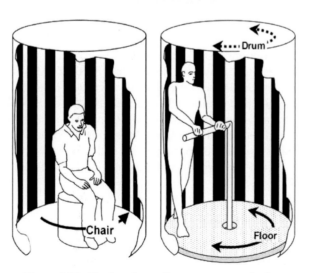

Figure 6.12. Vection drum. Image courtesy: Author.

$60°/s$. Following each 60-second exposure increment you will report MS symptoms (dizziness, headache, etc.) on a 1-to-5 scale. Following a 60-second rest session you will be exposed to vection again, for a total of five 1-minute exposures, or until you can go on no longer without becoming very sick!

2.2   Preflight Adaptation Facility

The second part of this module will take place in the Preflight Adaptation Facility (PAF) at Ames Research Center (ARC), where you will have the opportunity to experience a variety of visual–vestibular stimulus conditions which will help you recognize MS symptoms. The rationale behind you training in this facility is associated with the ability of humans to adapt to different environments. By repeatedly moving back and forth from a normal environment to a motion-provoking environment, your brain will be forced to utilize different sensory control programs. After repeating this procedure you will become quickly adapted to either environment and this training should help you to adapt prior to your flight.

At ARC you will also be trained to use the Device for Orientation and Motion Environments (DOME), which tilts and moves what you see while you remain motionless. The DOME consists of a 3.6-meter diameter spherical dome virtual reality (VR) system, the inner surface of which serves as a projection surface on which the observer views an imaginary environment while seated in a rotating chair which moves at speeds of $120°/s$ to $200°/s$. During this training module you will be exposed to the virtual environment of your operator's vehicle, in which you will perform various navigation-training sessions.

3   AUTOGENIC FEEDBACK TRAINING

3.1   AFT theory

Autogenic feedback training (AFT) is probably the most effective countermeasure that has been developed to counter the deleterious effects of SMS. It is a system that has been developed by Dr. Patricia Cowings of ARC and consists of a straightforward training procedure that combines the application of biofeedback and autogenic therapy (an acquired self-regulation technique) in controlling some SMS symptoms, such as vomiting and nausea. The 6-hour Autogenic Feedback Training Exercise (AFTE) program that you will complete at ARC will improve your motion tolerance by up to 85% and should further reduce the chances of SMS affecting your expensive orbital vacation!

3.2   AFT training

The principle of the AFTE training course is to teach you to voluntarily control several of your physiological responses using a combination of physiological and perceptual training techniques (such as autogenic therapy and biofeedback) that are

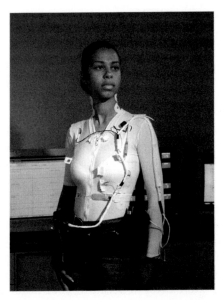

Figure 6.13. Components of the Autogenic Feedback System. Image courtesy: NASA. Source: *http://human-factors.arc.nasa.gov/ihh/psychophysio/technology.afte.html*

designed to relax your physiological profile and thereby reduce your reaction to stress. For example, autogenic therapy is a self-regulatory technique that has proved to be effective in modifying the autonomic response to stress by using self-suggestion exercises designed to induce bodily sensations that are highly correlated with specific physiological responses, such as peripheral vasodilation. You will find that after practicing these self-suggestion exercises you will be able to relax your physiological response to stress, thereby reducing the severity of any SMS symptoms. The biofeedback component of the AFTE course will inform you of sensory information concerning physiological responses to stressors by presenting you with your heart rate and/or blood pressure on a digital panel meter, thereby allowing you to recognize physiological changes associated with motion stimulation.

3.3   AFT system

During your training, you will wear the Autogenic Feedback System-2 (Figure 6.13), a biomedical instrumentation package originally designed and developed at ARC for use during the Spacelab-J (STS-47) mission. The system is a physiological monitoring and feedback system that consists of a shirt fitted with transducers, signal-conditioning amplifiers, a microcontroller, and a wrist-worn feedback display. You will wear this system when conducting your training with the Autogenic Clinical Lab System (ACLS), a PC-based physiological monitoring and training system which utilizes user-interactive software that directly measures and displays your physiological responses in real time.

The monitoring of your physiological responses is achieved by placing sensors and transducers at various locations on your body, and the information from these sensors is then displayed on a wrist-worn display unit. By using this system, you will be able to monitor the following parameters:

a. Blood volume pulse (BVP). A tiny infrared sensor worn on the small finger of your left hand will detect changes in the blood vessel volume of your hand.
b. Skin temperature. Another sensor mounted within the same unit as the BVP sensor will measure skin temperature.
c. Skin conductance level (SCL). Electrodes mounted on your left wrist will monitor changes in the electrical conductivity of your skin.
d. Respiration. Piezoelectric film strapped across your diaphragm will measure the range and frequency of respiratory cycles.
d. Electrocardiography (ECG). Electrodes will be placed on your chest to monitor electrical impulses from your heart.
f. Acceleration. An accelerometer attached to your headband will measure your head's motion along three axes.

4   VIRTUAL REALITY AND GENERIC PREFLIGHT VISUAL ORIENTATION AND NAVIGATION TRAINING

4.1   Introduction

A key component of this module is your opportunity to use synthetic environment training. The Simulated Intra Vehicular Activity System (SIVAS) you will be using during this training module uses data derived from video of your operator's vehicle on orbit that will allow you to view the vehicle interior with real geometry and lighting consistent with on-orbit activities.

4.2   Virtual reorientation illusion/intravehicular (IVA) training

Another objective of IVA training is to assist you in maintaining your spatial orientation during your stay on orbit. When determining orientation in a 1 G Earth environment, your brain utilizes input from a number of sensory channels, predominantly proprioception and vision [15]. Once on orbit, however, your proprioceptive sense often provides illusory information which means that you will need to rely much more on vision to maintain your spatial orientation, since inner ear cues will no longer signal the direction of "down" [13, 14]. One of the illusions you will encounter is the natural tendency to assume that the surface beneath your feet is the floor and the perception that the "walls", "ceiling", and "floors" will often exchange subjective identities! Also, when viewing a fellow crewmember "upside down", you may often feel upside down yourself due to the subconscious assumption carried over from life on Earth that people are usually upright. You may perceive that you will feel continuously inverted regardless of your orientation in the spacecraft due to the effect of fluid shifts and how receptors in your body interpret these inputs. This inversion is an artificial impression that is termed a "visual reorientation illusion" (VRI) and may cause you to experience symptoms of SMS, especially during your first few days of microgravity. These VRIs may also cause you to experience navigation difficulties when moving around the spacecraft. For example, you may find it difficult to remain oriented when traveling from one compartment to the next, as some compartments are not co-aligned. To reduce this difficulty you may be able to use landmarks within the spacecraft as frames of reference, and after a while, and with some practice, you will probably begin to visualize the three-dimensional relationships among the compartments and be able to traverse between them instinctively.

The importance of VRIs not only causes problems for new astronauts, but also represents a very real operational concern in the event of an emergency evacuation or explosive decompression. This is one of the reasons that preflight visual orientation as a countermeasure is utilized as a part of your training and why many of your responses to emergencies inflight are conducted using this training tool.

4.3   Virtual Environment Generator training

After completing your SIVAS training you will have the opportunity to be trained in the use of the Virtual Environment Generator (VEG), a VR system that can simulate certain aspects of microgravity and serve as a countermeasure for SMS and spatial disorientation (SD). Although our understanding of how the human sense of direction is neurally coded in microgravity is incomplete, research has suggested that the use of a VR system may help in reducing SD. The VEG you will use comprises a head-mounted (Figure 6.14) display (HMD), the position and orientation of which commands a computer to generate a scene that corresponds to the position and orientation of your head. This synthetic presence will allow you to move around in the artificial world of your operator's spacecraft.

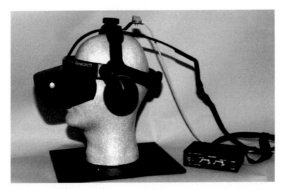

Figure 6.14. Virtual Environment Generator head-mounted device. Image courtesy: piSight. Source: *www.vrealities.com/pisight.html*

When you don the VEG equipment you will be presented with an image of the interior of your operator's spacecraft and a space-stabilized virtual control panel with an image of your hand in the HMD. As you move your hand the virtual hand will also move. Collision detection software in the graphics computer [2, 3, 12] will detect when your hand penetrates the virtual control panel, enabling you to interact with the virtual switches or objects to control events within the spacecraft.

You will also be able to manipulate objects in the virtual spacecraft, and be able to experience resistance to movement, texture, mass, and compressibility thanks to the haptic (tactile) and force feedback [2, 3] systems. To help you in the virtual spacecraft, the system has been designed to provide auditory cues when an object is grasped or dropped, or when a virtual switch is operated. This synthesis of visual and auditory cues will augment the visual information presented to you and thereby enhance your performance within the spacecraft.

Your operator will ensure that database compression techniques will result in the virtual spacecraft containing all objects, effect of human behavior, and effect of collision for real-time operation, which means that no matter how fast you move through the environment, you will experience no visual lags. The real-time operation will result in you being able to experience the high degree of realism and interactivity necessary to allow you to perform tasks necessary for training.

5   PARABOLIC FLIGHT

Microgravity can be simulated when an aircraft flies a Keplerian trajectory, or parabolic flight profile that requires the aircraft to climb rapidly at a 45° angle and then follow a path called a parabola (Figure 6.15). At the apex of the parabola

Figure 6.15. European Space Agency's Airbus performing a parabolic flight maneuver. Image courtesy: ESA. Source: *http://www.spaceflight.esa.int/users/index.cfm?act = default.pae&level = 11&pae = 1048*

the aircraft produces a near-zero-G effect (1×10^{-3} g) for between 20 and 28 seconds just as the aircraft achieves 9,500 m of the 10,000 m ascent (termed *pullup*) before it slows. The aircraft then traces a parabola (pushover), descending rapidly at a 30°–45° angle (termed *pushout*) to 7,300 m.

The acceleration forces produce approximately two times normal gravity (2 G) during the "pullup" and "pullout" phases of the flight, and between these phases is an intermediate phase, termed the "pushover" that occurs at the apogee, generating a zero-G environment with less than 1% of Earth's gravity. Due to the gut-wrenching sensations produced during parabolic training flights the aircraft has earned the nickname "vomit comet".

During your training you will fly three parabolic flight increments, during which you will fly more than 50 parabolas. Your parabolic flight indoctrination will take place onboard a specially modified Boeing 727-200, named G-Force-One™, which is an FAA-approved cargo aircraft designed to conduct parabolic flight maneuvers. The interior is divided into two zones: the rear zone designated the Seating Zone, and the forward zone (which is approximately 20 m long) designated the Floating Zone, the floor of which is covered with passenger-friendly, special, energy-absorbing Ensolite padding. Just like a regular airline, G-Force-One™ carries a crew that includes a captain, first officer, and flight engineer, and just like a regular airline, there is a flight attendant who is available to assist any crewmember who might feel sick.

Your microgravity indoctrination will gradually expose to you reduced gravity environments, beginning with 15 Martian parabolas in the second week that will provide you with one-third gravity, progressing to 15 Lunar parabolas (one-sixth G) in the fourth week, and finally 20 zero-G parabolas in the final week of training.

6   INCREMENTAL VELOCITY AND UNUSUAL ATTITUDE TRAINING

The purpose of the two sessions scheduled for this component of your training is to indoctrinate your body to the spatial disorientation and G forces you will experience during launch, re-entry, on-orbit, and in the event of an emergency egress. The course combines actual aerobatics with ground school, which will provide you with computer-animated enactments of the maneuvers you will be exposed to during your flight in the P51 Mustang. During the actual flights, most of the upsets and unusual maneuvers will be sudden and unexpected, and during most of the flight you will feel as if you are riding a rollercoaster that is slightly out of control!

P51 Mustang unusual attitude and G training

The program will begin with a safety briefing and a short indoctrination on how to wear a parachute, although by this stage you will have completed your accelerated freefall (AFF) course, so this will be familiar to you. Next will be a review of the emergency procedures. Because you will spend much of the 40-minute flight either upside down, looping or spinning, you will need to empty your pockets of cellphones, pens, and pencils. As the passenger, you will sit in the front seat and the pilot will sit behind you. Once you are comfortably seated you will buckle your five-point flight harness, which will be checked by the pilot before he climbs into his seat, after which he will conduct a brief cockpit orientation.

Shortly after taking off, the pilot will put the P51 into a sharp climb before leveling out at the practice altitude of 3,000 meters. The pilot will perform two consecutive 360° turns, one to the right and one to the left, at a shallow bank angle followed by two more turns at increasing bank angles and increasing speed. At the end of the fourth turn the pilot will flip the plane inverted and fly the P51 in this orientation for a while before flipping the plane right way up and entering a spiral dive. After bottoming out at 1,000 meters the P51 will climb steeply back to 3,000 meters at a progressively increasing angle of attack. At 3,000 meters you will feel the aircraft begin to shake as it approaches its stall speed, followed shortly after by the very real sensation that the aircraft is falling. After falling 1,000 meters the pilot will pitch the nose down into a sharp dive that will rapidly become a spin. The P51 will now be flying directly at the ground spinning faster and faster as the pilot calmly explains to you how he will recover the aircraft. A few more spins and the P51 will be flying wings level and the pilot will prepare for landing.

Supersonic flight

Your supersonic indoctrination ride will begin in the fitting room at the air force base, where an advanced technology anti-G suit (ATAGS), oxygen mask, and helmet will be waiting for you. Since you will have completed the high-altitude indoctrination and G-tolerance training course earlier, these items will be familiar to you. After suiting up you will receive a short indoctrination into the use of the ejection seat in which you will be sitting during your flight. Your pilot will then deliver a preflight briefing and you will proceed onto the apron, where a T-38 will be waiting to take you on your supersonic ride. If you think you might get airsick, now would be a good time to pop some pills!

The T-38 is a twin-engine, high-altitude, supersonic jet trainer, used by career astronauts as a means of maintaining flight currency and for traveling between Houston and Cape Canaveral. It also serves a role as a chase plane during Space Shuttle landings.

The ground crews will complete their work as you and the pilot climb into the pressurized cockpit where the two of you will sit in tandem on rocket-powered ejection seats. Since you will be flying at high altitude you will need to wear a mask that provides you with a 100% positive pressure oxygen source, but because you will be flying below the Armstrong Line your flight suit will not be a full-body pressure suit. Minutes later the T-38 will tear down the runway and take off, afterburners kicking in as the nose rises sharply. In less than a minute you will already be at your exercise altitude of 12,000 meters.

This will probably be your first opportunity to ride in a supersonic jet, an experience that comparatively will be a real kick in the rear compared with the P51. The training syllabus has provision for each crewmember to ride in the backseat for two rides, each lasting 45 minutes. Your pilot will warm up by performing some basic maneuvers, which will include some loops and basic acrobatics, similar to those you experienced in the P51. The difference in these maneuvers is that they will be performed at 500 knots! As you approach the practice area the pilot will point out some ground references so you can get a feel for the speed and provide you with some situational awareness. The pilot will then give you a "thumbs up" and once you return the signal the ride will begin. The first maneuver the pilot will perform will be the Split S, one of the oldest combat maneuvers, in which the pilot rolls his aircraft inverted and then executes a half-loop so at the end of the maneuver the aircraft is flying in the opposite direction. After the Split S the pilot will perform an aerial maneuver known as the Immelman Turn, which is a half-loop followed by a half-roll. To perform the Immelman Turn, the pilot will accelerate to sufficient airspeed before pulling the aircraft into a climb, continuing to pull back on the controls as the aircraft climbs. As the aircraft passes over the point at which the climb commenced the aircraft will be inverted and a half-loop will have been executed. At the top of the loop the pilot will execute another half-roll to regain the normal, upright orientation. On completion of the maneuver the aircraft will be at a higher altitude and will be flying in the opposite direction!

These first two maneuvers will constitute the warmup period!

Once you have recovered your stomach and you assure the pilot that you really do want to continue, the pilot will execute the Flat Scissors and the Rolling Scissors, which will be your first exposure to significant incremental G loading being imposed on your body. The Flat Scissors involves the pilot pulling the aircraft into a tight turn by pulling back on the control stick, which causes a sustained G loading. The tighter the turn, the higher the G loading. Since the turn is performed in one plane, or an imaginary flat surface, the maneuver is known as the Flat Scissors, whereas the Rolling Scissors maneuver is somewhat different, since it involves the pilot putting the aircraft into a dive followed by a pullup, which is known as a "zoom" climb. After the dive and pullup phases of the maneuver, the pilot will put the aircraft into an aileron roll, which involves rolling the aircraft along its longitudinal axis (the fuselage), and follow this with a barrel roll, which involves rolling and looping motions, climbing and descending and turning while the aircraft rolls around its longitudinal axis. By now, apart from feeling exhausted, you will probably be feeling more than a little discombobulated due partly to the effect of all the maneuvers on your situational awareness.

The experience is over much too soon, as you land back on the runway. You have just flown Mach 1.2, a speed that is less than one-tenth as fast as you will be flying on your trip to orbit, albeit without the aerobatic maneuvers.

References

[1] Brandt, T.; Dichgans, J.; and Koenig, E. Differential effects of central nervous peripheral vision on egocentric and exocentric motion perception. *Experimental Brain Research*, **16**, 476–491 (1973).

[2] Cater, J.P.; and Huffman, S.D. Use of Remote Access Virtual Environment Network (RAVEN) for Coordinated IVA–EVA Astronaut Training and Evaluation. *Presence: Teleoperators and Virtual Environments*, **4**(2), 103–109 (Spring 1995).

[3] Chung, J.; Harris, M.; Brooks, F.; Kelly, M.T.; Hughes, J.W.; Ouh-young, M.; Cheung, C.; Holloway, R.L.; and Pique M. *Exploring virtual worlds with head-mounted displays, non-holographic 3-dimensional display technologies, Los Angeles, January 15–20, 1989.*

[4] Dichgans, J.; and Brandt, T. Optokinetic motion sickness and pseudo-Coriolis effects induced by moving stimuli. *Acta Otolaryngology*, **76**, 339–348 (1973).

[5] Dichgans, J.; and Brandt, T. Visual vestibular interaction: Effects on self-motion perception and in postural control. In: R. Held; H. Leibowitz; and H.L. Teuber (Eds.), *Handbook of Sensory Physiology*, Vol. 8, *Perception*. Springer-Verlag (1978), pp. 755–804.

[6] Diener, H.C.; Wist, E.R.; Dichgans, J.; and Brandt, T. The spatial frequency effects on perceived velocity. *Vision Res.*, **16**, 169–176 (1976).

[7] Hu, S.; Davis, M.S.; Klose, A.H.; Zabinsky, E.M.; Meux, S.P.; Jacobsen, H.A.; Westfall, J.M.; and Gruber, M.B. Effects of Spatial Frequency of a Vertically Striped Rotating Drum on Vection-Induced Motion Sickness. *Aviation Space and Environmental Medicine*, **68**, 306–311 (1997).

[8] Hu, S.; Glaser, K.M.; Hoffman, T.S.; Stanton, T.M.; and Gruber, M.B. Motion sickness susceptibility to optokinetic rotation correlates to past history of motion sickness. *Aviation Space and Environmental Medicine*, **67**, 320–324 (1996).

[9] Hu, S.; Grant, W.F.; Stern, R.M.; and Koch, K.L. Motion sickness severity and physiological correlates during repeated exposures to a rotating optokinetic drum. *Aviation Space and Environmental Medicine*, **62**, 308–314 (1991).

[10] Lackner, J.R.; and Graybiel, A. Some influences of vision on susceptibility to motion sickness. *Aviation Space and Environmental Medicine*, **50**, 1122–1125 (1979).

[11] Reason, J.T. Motion sickness adaptation: A neural mismatch model. *Journal of the Royal Society of Medicine*, **71**, 819–829 (1978).

[12] Rebo, R.K.; and Amburn, P. A helmet-mounted environment display system. In: Helmet-Mounted Displays. *SPIE Proceedings*, **1116**, 80–84 (1989).

[13] Slater, M.; Steed, A.; McCarthy, J.; and Maringelli, F. The Influence of Body Movement on Subjective Presence in Virtual Environments. *Human Factors*, **40**(3), 469–477 (1998).

[14] Stanney, K.; Mourant, R.; and Kennedy, R.S. Human factors issues in virtual environments: A review of the literature. *Presence*, **7**, 327–351 (1998).

[15] Stanney, K. *Handbook of Virtual Environments: Design, Implementation and Applications.* Lawrence Erlbaum Associates.

Module 6 Onboard systems orientation and generic vehicle orientation

Contents

1 ONBOARD SYSTEMS ORIENTATION

Professional astronauts are required to have a comprehensive understanding of all systems and subsystems onboard their vehicle and habitat. They are also required to understand the relationships between these systems and to be able to identify major hardware components, state their function, identify intrasystem and intersystem interfaces, and describe the capabilities of each. Such a rigorous level of knowledge is not a requirement for orbital spaceflight participants, although a familiarity with the vehicle's primary systems is necessary as you will be required to assist in routine on-orbit maintenance tasks and must be proficient in reacting to onboard emergencies such as fire and decompression.

As an introduction to the primary onboard systems was provided on pp. 57–58 there is no need to repeat this information here. Instead, supplementary functions for some systems are described due to the greater complexity of these systems in an orbiting habitat.

1.1   Environmental Control and Life Support System

Overview

The Environmental Control and Life Support System (ECLSS) performs a number of vital functions onboard your habitat such as supplying air, water, and food. It also maintains temperature and pressure as well as shielding you from external influences such as radiation. The subject matter expert for the ECLSS will always be the flight engineer who must be familiar with each ECLSS component, which includes such systems as the Air Revitalization System (ARS), the Atmosphere Revitalization Pressure Control System (ARPCS), and the Active Thermal Control System (ATCS), each of which interact to provide a habitable environment for crewmembers.

Air Revitalization System

The habitat contains several independent air loops that circulate habitat pressure atmosphere. These loops constitute the ARS, which is responsible for circulating air inside the vehicle/habitat, ensuring humidity remains between 30% and 75%, ensuring carbon dioxide and carbon monoxide levels remain non-toxic, that temperature and ventilation is regulated, and that the habitat's avionics and electronics are cooled.

As the air circulates it collects heat, moisture, carbon dioxide, and debris before being drawn through a cabin loop and filter by a cabin fan which ducts the air to lithium hydroxide canisters which remove carbon dioxide and trace contaminants. The interval at which canisters are exchanged is a function of the number of crew and the size of the vehicle/habitat, but normally they will be exchanged every 12 hours. Once the air has passed through the lithium hydroxide canisters it goes through a heat exchanger and is cooled. At this stage, any water that is in the air is separated by a humidity separator fan which routes water to a waste water tank and the air is returned to the cabin. The frequency with which air is renewed depends on the size of the vehicle/habitat, but usually a cabin air change will occur between six and eight times an hour.

Atmosphere Revitalization Pressure Control System

The Atmosphere Revitalization Pressure Control System (ARPCS) ensures that the habitat air pressure is maintained at a pressure of 14.7 psia and that the partial pressures of oxygen and nitrogen are maintained within nominal levels. Since oxygen constitutes 20% of the air mixture, the partial pressure of oxygen must be maintained between 2.95 psia and 3.45 psia, whereas the partial pressure of nitrogen, which makes up 80% of the air, must be maintained at about 11.5 psia. Several specialized cryogenic oxygen tanks contain the source of the habitat's oxygen whereas the nitrogen is contained in several nitrogen cylinders. An average of 1.76 pounds of oxygen is used per crewmember per day, but these numbers to not take into account the normal loss of habitat gas to space and the amount lost to metabolic usage. Due to the different volumes of each orbiting habitat it is not possible to estimate how much nitrogen and oxygen is lost in this manner, but this information will be available in your operator's habitat reference guide.

The oxygen and nitrogen supply systems are controlled by the ARPCS that regulates the release of the gases by a system of check valves, inlet valves, relief valves, supply valves, sensors, control switches, and talkback systems. A description of the function of this system is beyond the scope of this module and you will not be expected to have an understanding of how the system works. However, you will need to know how to react if the system detects pressures outside nominal parameters.

If habitat pressure falls below 14.0 psia or rises above 15.4 psia, or if oxygen partial pressure falls below 2.8 psia or rises above 3.6 psia, then the Master Alarm will sound and red caution lights will illuminate on the ARPCS panel. Whichever crewmember is closest to the ARPCS panel will be responsible for dealing with the emergency, which will mean that you will need to know where the overpressure and negative relief valves are located. For example, if the emergency is an overpressurization situation you will need to activate the habitat relief switch which will in turn activate a motor-operated valve designed to relieve pressure through venting. If the emergency is a low-pressure alarm you will need to activate the negative pressure relief valves that will cause a flow of ambient pressure into the habitat.

Active Thermal Control System

The Active Thermal Control System (ACTS) component of the ECLSS is responsible for heat rejection, a function it achieves by the use of cold-plate networks, coolant loops, liquid heat exchangers, and various other heat sink systems that reject heat outside the habitat. The habitat has a large number of electronic units and systems that generate heat, and it is important that the heat sink systems are not overloaded, although if the capacity of the heat sink units is exceeded the habitat can activate a flash evaporator that is designed to meet excess heat rejection requirements for short periods.

Smoke and fire detection and suppression

A description of each item of firefighting equipment and the fire response procedures that will be employed by your operator is beyond the scope of this publication, but given the obvious potentially grave implications of such an event it is appropriate to review some generic firefighting capabilities and systems.

Smoke and fire detection and suppression capabilities are provided throughout the vehicle and orbital habitat by means of ionization detection elements that provide information concerning smoke concentration levels to the performance-monitoring general purpose computer (GPC). If the GPC detects an abnormal concentration of smoke the master alarm red lights will activate and the general alarm will sound throughout the vehicle and/or habitat.

Fire suppression in the vehicle and habitat will be achieved by the release of freon-1301 (bromotrifluoromethane), which may be released automatically by the GPS or manually by means of a push button. In addition to the fitted systems onboard, the crew can fight fires using portable, halon-1301 (monobromotrifluoromethane) handheld fire extinguishers that are located at various positions in the vehicle. There are also portable fire extinguishers (PFEs) which contain sufficient carbon dioxide to significantly reduce the concentration of oxygen in a confined space such as a habitat compartment, a feature that has implications for you if faced with fighting a fire. The operation of a PFE comprises a simple sequence of events that require you to remove the locking pin and depress the trigger. These actions will become intimately familiar to you as you conduct what will seem like endless fire drills in your vehicle and habitat mockups.

During the drills you will use the same PFEs that will be installed onboard the vehicle and habitat and also have the opportunity to become familiar with the Portable Breathing Apparatus (PBA), a system you will need to use when fighting most fires onboard. The PBA is a space-modified breathing mask similar to the ones that terrestrial firefighters use and comprises a mask to which is attached a 1.8 m hose that has a quick disconnect attached at one end. The disconnect end of the hose is simply plugged into a small oxygen cylinder that supplies sufficient oxygen for 15 minutes.

1.2   Operational communications

Communication between the vehicle/habitat and the MCC will probably be sent via a domestic communications satellite on a shared-time basis, which means that you will have limited time for communicating with family and friends. Direct signals from the MCC to the vehicle are referred to as *uplinks*, and signals from the vehicle to the MCC are referred to as *downlinks*, each of which are sent using the S-band portion of the radio frequency (RF) spectrum of 1,700 MHz to 2,300 MHz. Onboard the vehicle there is communication security (COMSEC) equipment that permits the encryption and decryption of confidential data, such as when a spaceflight participant confers with the flight surgeon and when you communicate with your family.

1.3   Guidance, navigation, and control

Guidance, navigation, and control (GNC) is effected using three simple steps. First, the guidance computer calculates the vehicle's location. Second, the navigation software tracks the vehicle's location, and third, flight control hardware transports the vehicle to the required location. During the launch and ascent phases the GNC assists in maintaining the vehicle's center of gravity, gimbaling the rocket engine clusters, and ensuring that thrust does not result in excess aerodynamic loading on the vehicle. The GNC system operates in either auto or manual mode, the latter requires the pilot to use the control stick to steer, whereas in automatic mode the vehicle's avionics software system permits the onboard computers to fly the vehicle. On orbit, the primary GNC tasks include achieving the correct position, velocity, and attitude necessary to accomplish the mission objective, which will usually be docking and undocking with the orbital habitat. Another task of the GNC system is to detect out-of-limit conditions, indications of which are provided by visual cues that consist of master alarm push button light indicators on the flight control panel.

2 GENERIC VEHICLE ORIENTATION

A part of your training will include equipment acquaintance sessions designed to introduce you to essential everyday items of equipment, a familiarity with which will be essential during your time onboard. Most of these sessions will combine a theoretical element that will be delivered using the Virtual Environment Generator (VEG), and a practical element that will be delivered in the vehicle and habitat mockups onsite. The following is a synopsis of the types of equipment that your operator will require you to train on.

2.1 Food Preparation System

Your habitat will be equipped with facilities for food preparation, dining, and stow-age that will be customized for each crewmember. You will have chosen your menu and food quantity when you completed your personal preference options module in Week 2 of your training program. This will have required you to choose menu food items that consist of three daily meals per day per crewmember, and pantry food items that consist of snack items and beverages. In addition to these two food categories you will also have chosen a contingency menu designed to provide you with food for up to 72 hours in the event of a de-orbiting delay.

In addition to choosing the food items you will be eating in space, you will also be able to express a preference for the way in which your food is prepared. For example, you will be able to choose between having fresh, irradiated, natural-form, thermo-stabilized, or rehydratable menu choices. The food preparation area in the habitat is designated a galley and consists of food warmers, food trays, and food system accessories designed to help you prepare your food. During the practical training phase for this module you will have the opportunity to practice using the various items of galley equipment in the habitat mockup facility. The instructors will demon-strate the use of the water dispenser and rehydration station to you and the method of operation in microgravity. You will learn that activities that we take largely for granted here on Earth often require much more thought and time. For example, the simple task of taking a drink will require you to insert a rehydratable beverage container into the rehydration station until the water dispenser needle penetrates the rubber septum on the rehydratable container, all the time ensuring that the specified amount of water is discharged into the container. Once you have managed that, you will need to mix the beverage and heat it, if required. The whole process, which would take perhaps 30 seconds on Earth, will probably take you 5 minutes or more in microgravity.

To prevent a mixup of food items, everything is color-coded, including acces-sories such as condiments, vitamins, gum, and candy. Due to the constraints of space in the habitat, the galley is designed for only one person at a time, so it is important that you become proficient in preparing your food, thus avoiding your fellow crewmembers being forced to waste valuable time waiting for you to finish! Generally, it takes one hour for one crewmember to prepare a meal, eat it, and to clean up.

2.2   Crew launch and entry suit and on-orbit clothing

During launch and re-entry, you will wear a pressure suit, which will consist of a helmet, communications cap, a pressure garment similar to an anti-G suit that fighter pilots wear, an anti-exposure layer, gloves, and boots. Over this Launch and Entry Suit (LES) you will wear escape equipment that consists of an emergency oxygen system, parachute harness, and parachute rig with an automatic opening device (AOD) and pilot chute, a life raft, 2 liters of emergency drinking water, a flotation device, and a survival vest pocket that will contain a search and rescue beacon, smoke flare, and sea dye marker. During the practical portion of this module you will have the opportunity to become familiar with the various items contained in your flight suit, and during the high-altitude indoctrination (HAI) portion of your training you will have the opportunity to test the integrity of your suit during your chamber ride to 25,000 meters.

During your personal preference options module in Week 2 you will have the opportunity to choose optional clothing and equipment items, such as a urine collection device, communications headset, emesis bag, kneeboard, watches, food and drink containers, and a flashlight. Your MRC module will also provide you with the opportunity to decide which clothing items you will use during your on-orbit stay. Most probably, your operator will provide you with the option of purchasing their own line of intravehicular activity (IVA) clothing such as shirts, sleep shorts, flight suits, and trousers.

Your U.S.$250,000 LES suit weighs 34 kg and is designed to protect you from sudden depressurization of the cabin during the launch and re-entry portions of the flights and against orthostatic intolerance following landing thanks to inflatable bladders in the legs. It will also protect you in the event of an emergency egress below an altitude of 18,000 meters and in the event of a water landing it will provide protection for up to 24 hours. The suit is a one-piece five-layer suit that includes a cover layer of fire-retardant Nomex. In addition to wearing the LES you will also don a full pressure helmet that has a faceplate that can open and close.

2.3   Sleeping quarters

The quality of your sleeping quarters will vary between operators but will probably consist of a sleeping bag, pillow, sleep restraints/adjustable straps, and a sleep station. How much space and therefore how much privacy you have will depend on the habitat construction and number of crewmembers, but you should expect fairly spartan quarters during the first few years of private orbital operations.

Your sleeping station will probably be located close to a porthole so you can go to sleep and wake up with a million dollar view. To ingress your station you will need to first unstow your personal sleep kit (PSK) from your locker. Next, decide whether you intend to sleep in a horizontal or vertical configuration, after which you can position your sleeping bag in the appropriate position, ensuring the adjustable straps are within reach. Next, make sure you have attached the pillow using the Velcro strips on the ends and slide into your sleeping bag and begin to restrain yourself using the elastic adjustable restraining straps. Finally, once you are ensconced within the confines of your sleeping bag and you have double-checked that you won't float away during the night, you can draw the privacy curtain, wrap the eye covers over your head and push your ear plugs into place so that the noise from the ventilation and ECLSS don't keep you awake. You will notice that inside your sleep station there is an air ventilation duct that acts as an air diffuser similar to the system you have in your car, which you can adjust using a push button screen. At the end of your sleep station there is a fluorescent fixture with a brightness control panel and an on/off switch.

2.4 Personal hygiene and general housekeeping

Your personal hygiene kit is a part of your initial kit issue and contains the necessary hygiene articles for brushing hair, shaving, nail care, tooth care, and general grooming. In addition to the kit, you will also be provided with two washcloths and two towels per day. If you require tissues, these are available from the tissue dispensers onboard your habitat, which dispense tissues that are highly absorbent.

Although you have paid more than $5 million for your one-week flight, you will still be expected to perform necessary housekeeping tasks that will require between 50 and 60 minutes of your time each day. The housekeeping chores will be divided up between crewmembers and a rota will be posted in the galley every day. The chores you can expect to perform include cleaning the galley, changing the air filters, trash disposal, exchanging lithium hydroxide canisters, and cleaning the waste management compartment. One of the more time-consuming chores is what is euphemistically called trash management operations (TMO), which will require you to collect and stow wet and dry trash, such as expended wipes and food containers. The management of wet trash is particularly important due to the potential for off-gassing, which will cause the contents of the wet trash bag to expand. To avoid this problem the wet trash bag must be connected to a vent in the Waste Collection System (WCS), which will ensure that any gas that does build up will be vented into space.

All the materials that you will need to perform these housekeeping tasks are stowed in lockers next to the galley. The equipment includes a vacuum cleaner, disposable gloves, general purpose wipes, and biocidal cleanser.

2.5 Photography and photographic equipment

Your standard issue camera is the Nikon D1, identical to the one used by NASA astronauts. It is a single-lens reflex digital camera with interchangeable lenses and is capable of single-frame shooting, capture preview, and record-and-review modes. It also allows for continuous shooting at a rate of 4.5 fps for a maximum of 21 shots. The camcorder that will be available onboard the habitat is the Sony HDW-700A, a high-definition camcorder that combines an HD color video camera head with a HDCAM portable videocassette recorder. Although you may find it a little heavy during your familiarization sessions, this is obviously a problem that you won't have to worry about during your stay on orbit! Also available to you during your trip is a motor-driven, single-reflex Hasselblad camera.

2.6   Restraint and mobility devices

In order to take good quality photos it will be necessary to use restraints that are provided throughout the habitat. These restraints will not only help you during your photography sessions but will also help you perform your galley duties safely. The most common restraint device is the foot loop, which is a cloth loop attached to the decks of your habitat. You will find foot loops installed near the portholes, galley, and at the workstations and if you need a foot loop you can simply install one yourself by unwrapping one and peeling off the adhesive before placing it in the desired location! Mobility aids such as handholds, footholds, and handrails are installed at various locations throughout the habitat and will allow you to move safely from one area to the next.

2.7   Equipment stowage

Each crewmember will be allocated two lockers, the dimensions of which are 30 cm by 45 cm by 60 cm. One locker is for personal items and small items of issued equipment, such as your cameras, whereas the other locker is for large issued items. Each locker contains two trays and dividers to provide a friction fit for microgravity retention. Other lockers are container modules that are designed to stow your LES and other launch equipment.

2.8   Exercise equipment

The longer your stay on orbit, the more time you will need to use this item of equipment. If you are staying for a week or less then you can expect to spend 30 to 40 minutes exercising per day. If you are staying for between 1 and 4 weeks your time spent exercising will be between 1 and 2 hours per day, which is why the treadmill is installed next to the biggest porthole of the habitat.

When you unstow the treadmill ensure you have all of its components, which should include a waist belt, two shoulder straps, four force cord extender hooks, and a heart rate monitor. Once positioned on the treadmill you will need to restrain yourself using the four force cords. Next, ensure your heart rate monitor is working and then decide how fast you would like to run by setting the speed control panel accordingly. As you run, the speed control panel will provide you with the time and distance run. Bear in mind that if you run for 90 minutes you will belong to a select group who have run once around the Earth!

2.9   Supply and waste water

The supply and waste water systems provide water for crew consumption and hygiene. The vehicle/habitat will have a number of supply water tanks containing potable water and one waste water tank. Due to obvious mass restrictions when flying into orbit, spacecraft do not stow water supplies onboard, instead they rely on fuel cell power plants to generate up to 12 kilograms of water per hour. To ensure that the water is drinkable the system is fitted with a pH sensor and a microbial filter that adds iodine to the water. Due to the effect of microgravity on any fluid, the water systems are pressurized using gaseous nitrogen. The water you will be using in the galley is ready chilled at a temperature of between 4°C and 8°C, whereas the ambient water temperature that is used for other purposes is between 17°C and 38°C.

2.10   Waste Collection System

The Waste Collection System (WCS) is located in the Waste Management Compartment (WMC); it is designed to collect and process biological waste generated by the crew. No doubt you have all heard the question often asked by school children: "How do you go to the toilet in space?" This module will explain everything.

Although the WCS appears similar to the toilet at home, there are several differences in its function and in its use which you must master before flying in space. During your practical module you will have the opportunity to use the WCS, usage of which is best described employing a step-by-step approach. In fact, to ensure no unfortunate events occur while a crewmember is inside the WMC, a checklist has been devised and is posted next to the toilet. It will provide instructions similar to those listed below.

Generic urine collection device instructions

- Lift funnel from holder.
- Remove lid from funnel
- Switch valve to "Open" position on funnel.
- Confirm "Separator" light is "On".
- Confirm airflow and position funnel are clear of body.
- Urinate.
- Twenty seconds following urination, close valve on funnel.
- Confirm "Separator" light is "off".
- Wipe funnel with washcloth and place in waste collection unit.
- Install lid on funnel.
- Place funnel on holder.

Generic urination/defecation instructions

- Remove lid from funnel.
- Leaving funnel in place, open plug valve.
- Confirm "Separator" light is "On".
- Confirm airflow.
- Affix solid waste collector insert on collector entrance and spread over seat.
- Lift funnel.
- Use solid/liquid waste collector unit.

- Lift seat, remove insert and place in solid waste receptacle.
- Wipe seat and funnel with bacterial wipe and place in waste compartment.
- Close lid of solid/liquid waste collector.
- Close valve on funnel.
- Confirm "Separator" light is "off".
- Replace lid on funnel.

Modules 7 and 8   Flight and emergency procedures

Most of your flight and emergency procedures will be taught using the Virtual Environment Generator (VEG) and animated pedagogical agents, which you will be able to see in stereoscopic 3D. Using the latest speech recognition software, you will also be able to talk to these agents as you conduct your training module. The lifelike characters will cohabit the vehicle during various simulated flight regimes to create a rich, face-to-face learning interactive environment. The agents will also be able to demonstrate complex tasks, employ locomotion and gesture to focus your attention on the most salient aspect of the task at hand. The agents will have a comprehensive knowledge of the domain in which the training takes place so they will be able to provide problem-solving advice to you and provide hints in case you have trouble solving a problem. Since your habitat and vehicle are complex environments, the virtual mockup and the animated agents will be an effective way of leading you around and helping you to learn where everything is before moving on to more complex tasks such as reacting to emergency situations. To assist you in learning as quickly as possible, the agents will have been designed with believability-enhancing behaviors that complement the advisory and explanatory behaviors that the agents will perform. For example, if you take too long in solving a particular problem, the agent assigned to you will exhibit "idle-time" behavior such as foot tapping. To create the illusion of life the agent will also exhibit full-body emotive behaviors such as expressive movements and visually complement the problem-solving advice they provide you. It will really feel like you are actually inside the vehicle or habitat. For example, your agent will scratch his head in wonderment when posing a rhetorical question. Since the agent is integrated into the Web-based delivery system that you have as a part of your computer-based training, your learning will be valuably enhanced.

This section describes a generic flight in a vertical takeoff vertical landing (VTOVL) spacecraft, similar to the New Shepard vehicle currently being developed by Jeff Bezos' company, Blue Origin. For the purposes of the fictional flight described here we will assume that the vehicle, the design and technical concept of which was inspired by NASA's DC-X, is a composite VTOVL base-first entry, single-stage-to-orbit spacecraft that consists of a propulsion section and a payload section that includes the cockpit and crew capsule.

The squat, bullet-shaped vehicle, which we will call "Pathfinder", will stand 15 m high and have a diameter of 7 meters. Propulsion will be supplied by a cluster of engines probably powered by high-test peroxide (HTP) and RP-1 kerosene for a total mass of approximately 54 tonnes. Similar to the design of the New Shepard vehicle, "Pathfinder" will feature four landing struts that will extend from the edges of the base of the propulsion section. Before describing what occurs during a launch we must first orient ourselves to the vehicle, a description of which is provided in the following paragraph.

The vehicle provides a crew station for a pilot and a flight engineer, each of whom is supported on the ground by a capsule communicator (CapCom), located at the Mission Control Center (MCC), whose job is to maintain real-time communications with the crew via a satellite link. The flight cockpit is similar to the layout of a corporate jet, the pilot sitting in the left seat and the flight engineer seated in the right. In front of the crew are duplicate instrumentation panels that provide all the information regarding the vehicle's performance. On the pilot's right side and the flight engineer's left side are centrally mounted engine throttles positioned forward of the three-axis reaction control system translation controller. The engine instrumentation layout includes strip gauges and emergency indicator panels located just below the quartz-tinted polycarbonate windows of the flight deck. On the left side of the pilot is a GPS flight director connected to a dual-differential GPS set, the batteries of which are independent of the vehicle's power supply, providing navigation redundancy in the event of an emergency situation resulting in loss of power. In an arrangement much like a commercial jet, the engines are controlled via a Full-Authority Digital Electronic Control System (FADEC), which, in the event of a loss of power, will permit manual control. Other important panels include the tachometer generator panel which provides the pilot with information regarding the RPM gauges, and the generator bus system that constitutes a system of relays that ensure each of the vehicle's systems is being powered. Next to this panel is a series of master caution and warning lights that indicate failure of any one of the four generators onboard. Below the tachometer generator panel is the turbo-alternator panel, which indicates to the pilot how electrical power is being supplied to the various electrical systems. The turbo-alternators use high-pressure, high-temperature gas from the combustion chamber to drive a turbine, which in turn drives up to four AC alternators, depending on the configuration of the electrical buses. Adjacent to this panel is the electrical system status board that indicates the amount of power coming through the Bus Power Unit (BPU), a device that ensures there are no spikes or overvoltages to the vehicle's systems. As with all other systems onboard, the BPU also has its own master and caution lights in the event of a short circuit. The information concerning the vehicle's hydraulic system is displayed on two identical panels, since there are two independent systems: one for odd-numbered engines, and one for even-numbered engines. The flaps, landing gear, and actuators, however, are powered by both systems working together.

Nominal flight procedures

You will have several opportunities to practice nominal and off-nominal procedures during your time in the systems trainer, during your VEG sessions, and during your CBT. Initially, this component of your training will focus on simply navigating in the vehicle and habitat, before proceeding on to routine emergency procedures and drills.

Toward the end of Week 5, shortly before your flight, you will have the opportunity to conduct several real-time nominal and off-nominal launches.

Preflight

- *Launch minus 2 hours.* The countdown begins with a call to stations by the Flight Director, a procedure that confirms each person of the launch team is in place at the MCC. Next, the backup flight system and the primary flight system computer software are checked and loaded onto Pathfinder's computer.
- *Launch minus 1 hour 45 minutes.* The flight crew begin stowing their gear, and flight engineers conduct a thorough inspection of the flight deck and exterior of Pathfinder. The work crew's module platforms are removed and loading preparations for the fueling commences. Once the launch pad is cleared of all personnel, the launch pad is declared closed and fueling commences.
- *Launch minus 1 hour 20 minutes.* Once fueling is complete the launch pad is declared open and the ground team and crew continue their preparations. The ground crew enter Pathfinder and switch on all flight control, navigation, and communications systems. Customized launch seats for each spaceflight partici-pant are installed and any payload destined for the orbital habitat is loaded onboard. During the fueling evolution, spaceflight participants start to suit up, assisted by launch personnel.
- *Launch minus 60 minutes.* A 60-minute countdown begins as ground team per-sonnel calibrate Pathfinder's inertial measurement units (IMUs). Meanwhile, in the MCC, personnel are communicating with nearby tracking stations and ensuring that tracking antennas are aligned for liftoff.
- *Launch minus 45 minutes.* The pilot and flight engineer conduct a walkaround of Pathfinder, a procedure similar to that which airline pilots perform before a flight. During their inspection of the vehicle, the pilot and flight engineer ensure the spacecraft has been properly serviced with the correct amounts of hydraulic fluid, oxygen, oil, and, of course, fuel! Once the pilot and flight engineer are satisfied that Pathfinder is spaceworthy, the pilot removes all duct plugs, locking pins, and probe covers and proceeds onto the flight deck. There, the pilot, with the assistance of the flight engineer, ensures all switches are in their normal shutdown position before conducting the cockpit inspection checklist. Some of the items checked include circuit breaker status, normal functioning of the pressurization and air-conditioning system, status of the fuel cell reactant valves, closing of the emergency dump valves, setting the fire handles, and checking the operation of the FADEC.
- *Launch minus 30 minutes.* The passengers will ingress Pathfinder while the pilot and flight engineer conduct their final preflight checks.
- *Launch minus 15 minutes.* The ingress hatch is closed and the Pathfinder's backup flight systems are transitioned to final launch configuration. Cabin vent valves are closed and the Flight Director receives a "go for launch" from the launch team. Once the "go for launch" verification is confirmed, the terminal countdown commences at the $T - 10$-minute mark.

- *Launch minus 5 minutes.* The MCC transmits a command that activates Pathfinder's operational instrument recorders, which store information relating to the ascent, on-orbit, and descent characteristics. The pilot and flight engineer complete the final checklist and are ready to start the engines, but before that they will run through some procedural checks with the MCC using a sequence of events that will be similar to the transcript below:

Pilot:	"Oxygen, set."
FE:	"Set."
MCC:	"Set."

Pilot:	"Regulator on, 100%."
FE:	"Regulator on, 100%."
MCC:	"Regulator on, 100%."

Pilot:	"Pressure suit, purged."
FE:	"Pressure suit, purged."
MCC:	"Pressure suit, purged."

Pilot:	"Hoses and connections, checked."
FE:	"Hoses and connections, checked."
MCC:	"Hoses and connections, checked."

Pilot:	"Helmet on, visor open."
FE:	"Helmet on, visor open."
MCC:	"Helmet on, visor open."

This checklist will continue for some time as the pilot and flight engineer cross-check with MCC that each vehicle system is running in its correct configuration for launch.

- *Launch minus 4 minutes.* The final communication exchange that will apply to you is the command to fasten your five-point harness and safety belt. Once you have confirmed to the pilot that you are strapped in, the pilot will begin to perform the final checks prior to starting main engines. You will hear the auxiliary power units start and will probably hear Pathfinder's hydraulic systems move the aerosurfaces through their range motions.
- *Launch minus 3 minutes.* Ground power transition occurs and Pathfinder's fuel cells switch to internal power. The pilot crosschecks normal operation of the radio, communications and navigation equipment, which include the radios, the radar altimeter, backup flight director, transponder, and launch azimuth. Once he is satisfied that these systems are operating normally, he retracts the drag flaps, sets the GPS radar altimeter to self-test, and sets the pressure altimeters.
- *Launch minus 2 minutes.* The propellant in the fuel tanks is brought to flight pressure. Concurrently, the MCC computers monitor hundreds of launch commit functions in case of any unusual event.
- *Launch minus 1 minute.* The vehicle's computers commence their terminal launch

sequence. You will know that the moment of truth is close at hand when you hear the following on the communication loop:

MCC: "Clear exhaust area."
FE: "Engine systems, checked. Engine systems, nominal."
MCC: "Exhaust area, clear."
FE: "Pressurization, high. Propellant boost pump switches, on. Main valves, open."
Pilot: "Number 3 and 6 engine turning. Throttle idle, depressing Start button."
FE: "Mode select switch Start."

- *Launch minus 45 seconds.* The pilot waits for a rise in combustion temperature and for two engines to stabilize in idle mode. Once he is satisfied that the engines are performing nominally, the pilot will repeat steps for two more until all engines are running nominally, at which point he will announce:

"Engine checks complete".

- *Launch minus 30 seconds.* With the engines running, you will be anticipating the moment of takeoff, but the pilot still has some final checks he must complete. The pilot will move his right hand to the booster engine throttles and push them slowly forward, at which point you will hear a noticeable increase in engine noise, until the pilot throttles back to idle, a sequence he will repeat for each cluster of engines, odd-numbered first, even-numbered second.
- *Launch minus 15 seconds.* As the pilot checks hydraulic pressures he will conduct a final briefing to you and your fellow crewmembers, an announcement that launch is imminent. Unlike a Shuttle launch there is no 10-second countdown as the pilot requests clearance from air traffic control in a similar manner to an airline pilot.
- *Launch minus 5 seconds.*

Pilot: "Transponder, normal. Flight controls, checked."
Pilot: "Oklahoma Traffic, this is Pathfinder ready for takeoff."
ATC: "Pathfinder, clear takeoff, Pad number three."
Pilot: "Clear takeoff."

Launch

After acknowledging receipt of the ATC clearance the pilot is committed to launch and will move all throttles forward while monitoring the alignment of Pathfinder and any trajectory deviation. Inside the MCC the mission-elapsed time (MET) is reset to zero and the mission event timer starts its count. During the first few seconds following takeoff, the flight engineer will monitor engine instruments, reporting any faults or abnormalities to the pilot. The takeoff will not be anything like the violence unleashed when the solid rocket boosters ignite for a Space Shuttle takeoff or the 4 G pushdown associated with a Soyuz launch. This takeoff will be little more than a nudge in your back, but as Pathfinder begins its ascent you will begin to feel as if you are gradually being pushed further and further into your seat as the velocity increases, until eventually, as vertical acceleration reaches its maximum level, you will be experiencing 3 G. The pilot maintains a running commentary about the events during the flight. Mach 1, the speed you experienced during your supersonic indoctrination training, comes and goes in an instant, followed almost immediately by the announcement that the vehicle is now travelling at Mach 5, then Mach 7. Mach 15! If you could lift your head far enough to see out of the window you would see blue rapidly becoming black as Pathfinder races towards orbit, but everything weighs three times as much as it does on Earth so the effort is too much.

On orbit

The first words you will hear confirming you have reached orbit will be:

Pilot:	"Throttles at cutoff."
FE:	"Altimeters, checked. Fuel cells, checked. FADEC, off."
Pilot:	"Drag flaps, closed. RCS, to manual."
FE:	"RCS valves, open."
Pilot:	"Unnecessary equipment, off."
FE:	"Unnecessary equipment, off."
Pilot:	"On-orbit checks, complete."
FE:	"On-orbit checks, complete."

The next stage of your journey is rendezvous. If you had been traveling in a Soyuz you would have had a two-day trip ahead of you, but a VTOVL launch aligns directly with the orbital plane of your habitat so the time between orbital insertion and rendezvous is very short. In fact, only a few minutes after orbital insertion you will hear the pilot and flight engineer run through a series of checks prior to conducting the rendezvous.

Pilot:	"Rendezvous radar, on."
FE:	"INS, GPS, checked. Rendezvous mode, on."
Pilot:	"RCS mode switch, on."
FE:	"Pressurization check, low."
Pilot:	"Rendezvous checks, complete."
FE:	"Rendezvous checks, complete."
Pilot:	"Flight. Rendezvous."

Once rendezvous checks have been completed there will be a short wait while the pilot aligns the vehicle with the habitat using some orbital aerobatics, probably flying a rendezvous pitch maneuver (sometimes called an RBAR pitch maneuver), developed by NASA engineers Steve Walker, Mark Schrock, and Jessica LoPresti after the Columbia disaster. At a distance of approximately 200 meters from the habitat Pathfinder will perform a slow 360° pitch, which allows personnel inside the habitat to perform a high-resolution photo survey of the underside of the vehicle for any damage to its thermal protection system. Once the habitat commander has confirmed to Pathfinder's pilot that they have captured the necessary images, the approval for docking will be authorized.

Pilot: "RCS switch."
FE: "RCS propellant pressures, checked."
Pilot: "Fuel cells."
FE: "Fuel cells, checked."
Pilot: "Docking collar, deployed."
FE: "Docking collar, deployed. Indicators, checked."

The pilot will then fly Pathfinder at an excruciatingly slow speed toward the habitat using a laser rangefinder and a targeting indicator to guide him. The moment of impact will be hardly noticeable, so the first indication you will have of a successful mating of the vehicle and the habitat will be via the communication loop.

Pilot: "Docking collar. Capture. Hard dock. Confirm good seal."
FE: "Docking collar. Hard dock. Confirm good seal."
Pilot: "Indicators, checked."
FE: "Indicators, checked."

After docking, leak checks will be performed and ingress activities will be prepared following a 20-minute safety briefing conducted by the pilot. Once the flight engineer is happy, the airlock fan will be bypassed and deactivated. The pilot will check with the habitat commander that the vehicle and habitat configuration, "the stack", is in the correct attitude and an attitude control handover procedure will be conducted that places the control of Pathfinder over to the habitat commander. Once the spaceflight participants have been transferred from Pathfinder to the habitat the vehicle will prepare to undock so it can de-orbit and return the passengers whose one-week trip in space has come to an end. For those disembarking, the first checks over the communications loop will be the following:

Pilot: "Oh two, set."

This initial command is the pilot ensuring that the pressure suit regulator is in the "On" position, the pressure suit is purged and all hoses, connections are checked. The pilot and flight engineer will then conduct a thorough orbital checkout of Pathfinder's systems that will be used during re-entry. This checkout will include checking communications and the status of the fuel cells, flight computer, avionics, hydraulic motors and hydraulic switching valves, the cycling of all aerosurfaces, radar altimeter, TACAN, and crew-dedicated displays.

Pilot: "Radio, on."
FE: "Fuel cells, on. Reactant valves, open. Cryo pressure, checked. Avionics bus, checked."

Pilot: "GPS and radar altimeter, test. Secondary GPS, on."
FE: "FADEC, on. INS, set. GPS, set. INS in Nav mode, check.GPS in Nav mode, check. Flight plan verified."
Pilot: "Flight plan verified. Lights, set. Throttles at cutoff. Drag flaps fully retracted."
FE: "Communication and navigational equipment, set."
Pilot: "Transponder, set. Downlink and uplink status verified. Autothrottle, off. RCS switch, on."
FE: "RCS valves, open. Collar umbilical, retracted."
Pilot: "Harness fasten, prepare for departure."

At this point you should have fastened your five-point safety harness and be preparing to wave goodbye to any friends you may be leaving behind on the habitat. After undocking, the pilot will conduct a ten-minute briefing prior to commencing the de-orbit burn that will align Pathfinder with its descent trajectory. You will know that you are approaching the landing site when you hear the radar altimeter being set, as this is normally done at an altitude of 1,500 meters.

Pilot: "Radar altimeter, set."
FE: "Pressurization, set. Propellant, on."
Pilot: "Pressure altimeters, set."
FE: "Descent mode, verified."
Pilot: "Descent checks, complete."

The pilot will maneuver using the RCS jets to orient the vehicle into the de-orbit attitude (retrograde) after which Pathfinder will fly in coast mode until the atmosphere and dynamic pressure buildup is reached at an altitude of 120,000 meters, a height referred to as the entry interface (EI). At this point the pilot will load de-orbit and entry flight software into the flight computer. This information is also sent to the MCC, which can, if necessary, input delta-state vectors to correct for any navigation errors that may occur during re-entry. Before EI the vehicle is still traveling at 7,500 meters per second, but, unlike the Shuttle, which at this point would be more than 8,000 kilometers from the landing site, you are less than a 160 kilometers away from home, thanks to the unique flying characteristics of a VTOVL vehicle.

Contingency procedures

The VTOVL vehicle you are riding into orbit provides emergency procedures throughout the ascent flight regime, unlike the Space Shuttle, which provides abort capabilities through only a narrow segment of the ascent profile. Also, unlike the Shuttle, for which one of the emergency procedures is to destroy the vehicle with explosive charges after takeoff, the VTOVL does not feature a remote self-destruct option.

In the course of your training you will receive instruction on how to cope with certain abort and contingency situations. The crew's chances of survival will be significantly increased if each crewmember has a thorough knowledge of their emergency duties in each contingency situation. If time and circumstances permit you will be notified by the pilot of an emergency situation via the communication loop. If a verbal warning is not possible it is likely that you will realize that an emergency is occurring by the sound of the klaxon and the panels lighting up like a keno board!

Emergencies may include circumstances such as engine shutdown, depressurization, throttle failure, engine overheating, gearbox failure, turbine damage, fire, electrical failure, or any one of dozens of other off-nominal events. Since specific emergencies require specific actions on the part of each crewmember, it is important to recognize the type of emergency that is occurring. The primary emergency situations and their associated alarms are as follows:

Ground evacuation:	One long sounding of the klaxon.
Action:	*Abandon vehicle.*
Emergency egress:	Verbal command *Bail out* will be given over the communication loop *three* times.
Action:	*Abandon vehicle* using emergency egress system.
Crash landing:	Alarm will ring three times.
Action:	Brace for impact
Depressurization:	A siren will sound continuously and a rotating blue light will illuminate.
Action:	Stay calm, as you will be wearing your pressure suit, which has sufficient air to keep you alive for four hours.

A description of the myriad emergency and contingency procedures that apply to each type of launcher is beyond the scope of this guidebook, but it is useful to describe the sequence of actions required in an emergency situation that are very similar for all launchers by virtue of economy and weight constraints.

Perhaps the most traumatic and harrowing emergency actions are those required to perform an egress from the vehicle. In this event, you will need to use the Crew Escape System (CES) that will probably be very similar to the one featured on the Shuttle, for reasons of economy and weight. The components of the CES include various pyrotechnic devices and an egress pole. The first stage of the egress procedure is to perform a controlled depressurization of the cabin, an event that is initiated by depressing a pyrotechnic vent valve. Generally, the higher the altitude at which venting occurs, the longer it will take for cabin pressure to equalize with ambient air, but even in the most extreme cases the length of time should not exceed one minute. Once depressurization is confirmed, the next step is to blow the crew ingress/ egress hatch by means of activating a series of linear-shaped pyrotechnic charges, which will sever the bolts attaching the hatch to the vehicle. Once the hatch is quite literally blown off the vehicle the egress pole must be removed from stowage and extended through the hatch opening. Now, it is time to clip the lanyard attached to the escape pole to your harness ring and slide out of the hatch and off the end of the pole, allowing gravity and the automatic opening device on your parachute to do the rest.

7

Commercial applications of space tourism

Space tourists, like their terrestrial counterparts, will use discretionary funds to travel and visit destinations that remove them from everyday life. Regardless of whether they pay for a suborbital flight or a trip to Titan, in each case spaceflight participants will be paying for the experience, an experience which will almost always require a significant infrastructure to support and which only the commercialization of space can accomplish.

Now that companies are finally retreating from the view that space tourism is science fiction, the potential of space commercialization may be fully realized. As infrastructure is put into place and more spacecraft come into use, the cost of accessing orbit will decrease to a level that space vacations will become financially viable for many middle-income people. As infrastructure develops, changes will be noticed in the comfort of spacecraft, resulting in space travel becoming more luxurious and less adventurous. Initially, however, the adventure rather than the cruise ship model of space tourism will be more attainable, although many space tourism advocates favor the latter and consider it necessary to attract the financial investment to sustain the interest of potential space tourists.

In this chapter, an overview of the future of the commercialization of space as it applies to space tourism is discussed. It is impossible to predict all the commercial applications that may eventually be fueled by the nascent space tourism industry. Hence, this chapter focuses only on those applications that will have the most immediate effect on the first generation of orbital spaceflight participants.

COMMERCIALIZATION TIMELINE

Initially, commercialization will take the form of suborbital flights, which will offer brief but intense experiences, allowing spaceflight participants to gain bragging rights. Within a few years of companies such as Virgin Galactic commencing suborbital operations, orbiting vacations will be offered, extending the period of microgravity from four minutes to several days or weeks. Orbital vacations will inevitably lead to orbiting space hotels, which may, initially at least, take the form of inflatable habitats such as Bigelow's BA330 module. Despite the significant planning and investment effort required to build an orbiting space hotel, there is sufficient promise of an economic return that it will be worth committing the funds. Orbiting hotels will surely be followed shortly after by space cruisers, which will probably require the largest increase in space infrastructure such as travel agents, food and beverage suppliers, and maintenance and docking facilities. Just as cruise ships on Earth offer a range of accommodations, space cruisers will no doubt offer similar grades of accommodations based on budgets and tastes. Initially, the only destinations will be the Moon and near Earth objects, but eventually, trips to Mars, Venus, and the outer planets will become the norm, at which point space tourism companies will begin to offer resort locations, as described in Chapter 8. Shortly after manned missions to Titan and Europa are accomplished, the colonization of the Solar System will begin in the form of orbital colonies in Earth orbit, followed shortly thereafter by colonies orbiting planets farther afield. Space colonization, like the real estate business on Earth, will represent the greatest commercial application in space and will require an infrastructure capable of supplying helium-3 for advanced fusion reactors, advanced solar power systems capable of supplying the needs of orbital colonies, and radiation protection.

Before these events can occur, however, the means of accessing orbit must be drastically reduced. Space advocates have identified several possible opportunities for the future commercial use of space, but until recently these opportunities depended on lowering the cost of transportation into space by several orders of magnitude. Now, with several suborbital companies and a few orbital companies within striking distance of opening up the final frontier to private entrepreneurs, the last significant barrier to further space development will disappear.

SUPPORT INFRASTRUCTURE

Based on the precedent of decades of commercial aviation experience it is probable that service prices will fall as passenger numbers rise, a situation that will inevitably give rise to a greater on-orbit infrastructure. Such an infrastructure will include an orbital accommodation industry, beginning initially with standard prefabricated modules not dissimilar to the International Space Station (ISS), but eventually developing to become large structures that will incorporate entertainment complexes and sports centers. An increase in passengers will also give rise to an increase in the demand for staff and for support infrastructure, such as service stations that will be located in each hotel orbit for the purpose of supplying water, oxygen, and hydrogen.

Although much has been written and predicted regarding orbiting hotels, this reality will only occur when flight prices are reduced to a level that permits thousands of passengers to visit orbit each year, a scenario that will probably not be realized for

As a commercial enterprise, space diving would seem to possess all the attributes to succeed in a world where extreme is the new norm when it comes to adventure sports. Better yet, for Clark and Tumlinson, their test pilots will be adrenaline-addicted space divers who will be paying for the privilege of risking their life, and in so doing, provide valuable data on how humans perform through the descent profile.

"Back toward the trailing edges then, to a small escape-hatch beside which was fastened a dull black ball . . . He gasped as the air rushed out into near-vacuum . . . He rolled the ball out onto the hatch, where he opened it: two hinged hemispheres, each heavily padded with molded composition resembling sponge rubber . . .

. . . He curled up into one half of the ball: the other half closed over him and locked. The hatch opened. Ball and closely-prisoned man plummeted downward."

Excerpt from the classic E.E. "Doc" Smith novel, *Triplanetary*

Diving from space

Space diving from the upper reaches of the atmosphere presents several dangers, which include thermal stress, increasing G, and speed, the latter being the least understood. The speed descent profile will require a space diver to enter the atmosphere at 4,000 km/h (2,500 mph) and slow down to the conventional skydiving speed of 192 km/h (120 mph). The thermal stress imposed on the space-diving suit will be in excess of 400°F, which, while representing a tricky engineering problem, is not insurmountable. The next challenge for the space diver will be the G loads, which will reach but not exceed 4.4 G. While these latter stresses may be survivable, the speed component may not, due to the absence of knowledge regarding the effects of Mach speeds on the human body. Kittinger, when performing his jump in 1960, reached an estimated subsonic speed of 1,025 km/h (641 mph), whereas the first space diver will become only the second transonic human, exceeding not only Mach 1, but also Mach 2 and Mach 3! The only precedent for a human surviving such a speed is SR-71 Blackbird pilot Bill Weaver, who survived a 78,000-foot, Mach-3.18 disintegration of his aircraft in 1966 without lasting damage. Weaver was undoubtedly very lucky as the shock waves that are produced in the Mach speed regime have the very real potential to simply rip limbs from the sockets as a result of an unstoppable flat spin. Conventional skydivers encounter flat spins but they can "push" against the air and correct accordingly, an option that will not be available to a space diver plummeting from the upper reaches of the atmosphere, where air density is much less. If they survive the flat spin, space divers may have to contend with the effects of "shock–shock" interaction, a phenomenon that may ultimately result in body fragmentation! All of which only serves to increase the appeal to those seeking the ultimate "rush"!

As so often has been the case when considering a possible future in space, the visions have always been based on boundless optimism, a problem perpetuated by government-funded space agencies since the dawn of the Space Age. Another problem is that the general public have been conditioned by the aforementioned space agencies to believe that it is impossible to conceive of access to space as inexpensive and popular. However, the fundamental disconnect between the $400 million cost of Space Shuttle flight and the tens of thousands of dollars it will cost to send a person into suborbit in the next five years has finally been broken by the success of SpaceShipOne. The commencement of Virgin Galactic operations in 2009 together with the concerted efforts mounted by other private space companies will inevitably lead to tourists embarking on the trips described in Chapter 8.

8

Advanced space tourism

LUNAR TOURISM

Living as a lunar tourist

The Moon is likely to be one of the most exotic and expensive tourist destinations of the latter half of the 21st century and the early part of the 22nd. For those embarking on a trip to the Moon your transfer flight will probably begin from an orbiting habitat in low Earth orbit (LEO) and will last between 3 to 5 days, depending on the launch window and propellant consumption. On arrival in lunar orbit you will be transferred to another orbiting module where you will be briefed by a tour guide representative and issued the equipment necessary for your surface stay. Alternatively, you may have chosen to remain in the module and observe the Moon from orbit, in which case you will need to become familiar with your home for the next few days.

Your living quarters will comprise a module containing single or double cabins placed around a central connection tunnel connected to another module housing a viewing platform. The third module will serve as accommodation for the flight engineer and crew and will be linked to the fourth module, which will consist of an airlock chamber enabling you to perform extravehicular activity (EVA). The several viewing windows of the habitat will serve an important dual function during your EVA, as it will allow you to maintain eye contact with those inside and thereby allay any feelings of claustrophobia.

If you have chosen the expensive surface excursion option you may decide to stay at one of the habitats located near the site of the lunar landings that will house not only fellow tourists but also scientists and engineers. The audaciously high design of your hotel will be a function of the Moon's one-sixth gravity, the windless environment, and tourist's requirement for spectacular views. Another unique design aspect will be the crew rescue vehicle (CRV) that will be docked to one of the hotel's connection nodes in the event of a solar flare that may require an emergency return to Earth. In addition to the lunar landing habitat you may wish to visit hotels located at the Moon's equator, which will provide you with views of the Earth directly over-head, or you may instead decide to move to higher latitudes to view the Earth closer to the horizon. Another option will be to visit a hotel located at the South Pole in an area of eternal sunlight. Regardless of where you view our planet it will appear four times as big as the Moon looks from Earth.

Lunar activities

For those with sporting inclinations, there will inevitably be several golf courses, although unlike the ones on Earth these will have emergency shacks located at each hole in case anyone has a problem with their spacesuit or to provide shelter in the event of a solar flare. In addition to playing golf it will be important for you to perform other exercises as gravity is just one-sixth as strong as it is on Earth. In fact, the longer your stay on the Moon the more important it will be for you to maintain a rigorous exercise regime, a factor that will cause lunar cities to ban most forms of mechanized transport to ensure tourists do not lose too much bone density.

When to go

The good news concerning lunar tourism is that it is available now, but only to the select few who have the required U.S.$100 million for the round-trip ticket, which, incidentally, does not include a surface stay. Space Adventures, a private space tourism company, is offering the ticket, and CEO Eric Anderson says he already has prospective spaceflight participants who have expressed interest and who can afford the ticket. The trip will use a modified Russian Soyuz TMA spacecraft and will feature the opportunity to view earthrise from lunar orbit and a view of the far side of the Moon from an altitude of 100 km. The Soyuz spacecraft, which is capable of carrying one pilot and two passengers, is the most reliable spacecraft in history and was originally designed to ferry astronauts to the Moon, although none of these missions occurred.

Preparations

If you do decide to travel to the Moon you will need to be very well prepared. Apart from the problems associated with bone demineralization that will afflict those tourists staying longer than a week, you will also need to be prepared for dealing with increased G loads. For example, during the trip you will experience 3 G during launch, 0 G during the 3-day stay in Earth orbit, and 3 G during translunar injection. During the landing phase you will again be subjected to 3 G, after which you will be exposed to one-sixth G during your lunar stay. Taking off from the Moon will impose 3 G, which is the same load that will be imposed on you during Earth injection. The heaviest G loads will be experienced during the aerobraking maneuver on return to

Earth when you can expect to tolerate up to 6 G. Finally, during the final phase you will experience another 2 G during landing on Earth.

VENUS

Venusian characteristics

Prior to 1960, thanks to science fiction writers such as Robert Heinlein, Venus was depicted as a warm oceanic world with islands covered in jungles and swamps. After a couple of probes landed on Venus, this version proved to be little more than an illusion. Instead of the paradise world envisioned by Heinlein, Venus was discovered to have an unimaginably thick carbon dioxide atmosphere, hardly any water, and several other disadvantages that immediately relegated it to the "no visit" list for humans, although it still continued to attract several probes. Although it may not seem an obvious choice as a tourist destination, there are some factors that may appeal to those interested in exotic locations. For example, the cloud deck over Venus is very high over the surface (49 km to 64 km/30 mi to 40 mi) compared with the cloud deck on Earth (8 km to 10 km/5 mi to 6 mi) and just below this cloud deck are panoramic views of the Veneran surface which can be enjoyed at temperatures and pressures similar to ones we find comfortable on Earth.

Traveling around Venus

The pseudo-ecosphere that covers our twin planet will be enjoyed by tourists traveling in aerostats that will use locally available hydrogen for buoyancy. An aerostat is simply a buoyant structure such as a balloon or blimp capable of remaining airborne indefinitely so long as its relative buoyancy is maintained. The aerostats will use lightweight composite materials and fabrics presently available, which would probably be mined *in situ* from the Venusian atmosphere. The tourist vehicle will comprise a pressurized crew compartment constructed of Kevlar carrying an inflatable gas envelope made of polybenzoxasole and would be dropped into the Venusian atmosphere together with a shuttle which would be used by the crew to return to orbit. For travel through the atmosphere the aerostat would use methane/oxygen-fueled propellers, but since landing in the event of an emergency will clearly not be an option, the aerostat will have backup power in the form of a methane/oxygen-fueled rocket engine.

Inevitably, one day tourists will want to descend to the surface so they can visit the planet's continents, explore its islands, and sail across rewatered oceanic basins, but the prospect of tourists becoming Venerans will not happen until the terraforming of Venus has occurred, a process that will involve a Herculean biosphere genesis project that may take centuries.

One option for especially wealthy tourists will be to combine visits to Venus with trips to Mars using what is known as a conjunction-class trajectory. In this scenario your spacecraft will first swing in toward Venus for a gravitational boost during which time you will have time for a short Venusian visit before continuing on to Mars for a two-month surface excursion before flying home.

MARS

"Light winds from the southeast in the early evening, becoming light winds from the east shortly after midnight. Maximum winds approximately 30 kph. High temperature will reach $-35°C$ and an overnight low of $-86°C$. Pressure will remain steady at 7.6 millibars. Radiation exposure assessed at moderate. Be sure to wear your shades."

Mars by the numbers

For the most part, Mars is freezing cold, with temperatures falling $-130°C$ at the poles. Its atmosphere, with a ground level pressure less than one-hundredth that of Earth's, consists predominantly of carbon dioxide and little oxygen and is not dense enough to trap much of the Sun's warmth or to shield the planet from the Sun's deadly ultraviolet radiation. However, these hostile conditions are tempered by some similarities with Earth. For example, a Martian day is only 37 minutes longer than a day on Earth, although a year on Mars lasts 687 days and there are 24 months, each of them 28 days long. Also, since Mars is tilted on its axis by $24°$, it experiences seasons similar to Earth.

Getting there

Outbound tourist traffic between the Earth and Mars will be a series of two-year peaks and troughs dictated by the proximity of the planets and the constraints imposed by minimum-energy launch windows, which occur at intervals of 2.135 years, or every 780 days. Like the outbound launch windows, inbound launch windows will also occur at intervals of 780 days.

What to see

Traveling around Mars will be relatively easy due to the planet's gravity being only one-third that of Earth's. This means that any backpacker will be able to carry three times as much as they can on Earth! Classic tourist destinations will be Olympus Mons (Figure 8.1) for those who are interested in volcanoes or enjoy mountaineering or skiing; Cydonia, for those interested in conspiracies; and Ares Fjord, site of Pathfinder and Sojourner, for those interested in space history. For those interested in the canyons of Mars the must-see destination will be Candor Chasma, located in the 2.5-mile-deep Valles Marineris canyon system. Winter enthusiasts will be attracted to Chasma Boreale in the north, with its residual icecap, whereas those interested in viewing those historic canals will want to visit Mangala Vallis near the equator. Unfortunately, traveling across the surface of Mars presents several

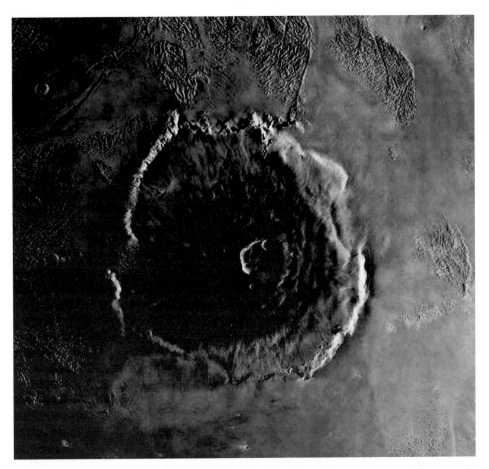

Figure 8.1. Olympus Mons. The tallest mountain in the Solar System and future tourist destination. Image courtesy: Science Fiction Fantasy Net. Source: *www.sff.net/people/ ckanderson/olympus.htm*

challenges, even for roving vehicles, although tourists will probably use these for short journeys of up to 50 mi/80 km. For longer journeys it will be necessary to rely on some unusual flying machines such as open-bottomed balloons and Marsplanes.

Traveling on Mars

For those tourists not concerned with speed, the balloon will be the preferred option. The design of an open-bottomed balloon is driven by the necessity for atmospheric gas to flow in and out, thereby permitting a daily rising and sinking. If a balloon were sealed then it would rise only to an altitude at which expansion of the internal skin would make its skin taut. By venting gas as required, Martian balloonists will be able to use wind currents at varying altitudes to travel in the desired direction.

For those tourists who need to get somewhere fast, the Marsplane will be the travel mode of choice. The first Marsplanes will look similar to a powered glider with a large wingspan of 20 metres, which will enable the aircraft to gain atmospheric lift in the anorexic Martian atmosphere. Due to the absence of airports, runways, and other aviation infrastructure in the early years of Martian tourism, early Marsplane designs will probably feature a catapult takeoff.

Excursions to Phobos and Deimos

Phobos and Deimos, which are most likely captured asteroids, circle the dry radiation-drenched world of Mars in orbits lying directly above the planet's equator. Phobos, the outermost moon, and Deimos, the innermost, are irregular cratered chunks of rock that are among the darkest bodies in the Solar System. Although there may not be much of interest to see on either moon, since Deimos is only a stone's throw away from Mars it is possible operators will locate a base there, as staying on this moon will shelter tourists from meteoric particle bombardment and high-energy solar proton events.

ASTEROIDS

If a two-year round trip to Mars or Venus sounds a little long, but you still want to stretch your space legs, you may want to consider a 60–90-day flight to a near Earth object (NEO) asteroid.

In 2007 NASA sent a robotic probe named Dawn on a mission to fly to asteroids Ceres and Vesta, each of which could one day be a destination for a future space tourist. Vesta is an asteroid that may implicate supernovas in the birth of the Solar System, whereas Ceres, measuring 905 km in diameter, is the largest object in the asteroid belt and is covered in an ice layer between 60 km and 120 km thick. An interesting aspect of a trip to the asteroid belt will be asteroid hopping, which will require the vehicle to undertake maneuvers from the orbit of one asteroid and fly to the orbit of another.

EUROPA

Any manned mission to the outer planets will be extremely challenging due to the distances, time, and technical considerations involved. Also, since planets such as Saturn and Jupiter are gas giants, they have no solid surface, so it will not be possible to land on them, although flying through the upper atmosphere will probably be possible with a specially designed craft that is several decades away from the drawing board. Of more interest to interplanetary space tourists will be the many moons that orbit the gas giants such as Io, Ganymede, Callisto, Europa, which orbit Jupiter, and Titan, which orbits Saturn. These moons are likely candidates for destinations that humans will one day walk on and eventually live on. However, even assuming that

Figure 8.2. Europa. Image courtesy: University of Arizona. Source: *http://pirlwww.lpl.arizona. edu/missions/Galileo/releases/6Mar_i25europa.htm*

there exists a means of transporting humans to Jupiter, the challenges posed by the deadly radiation near the planet will demand that substantial shielding be required as a part of the spacecraft's design.

Europan introduction

Europa (Figure 8.2) orbits 665,920 kilometers from Jupiter, whereas the Moon orbits 381,600 kilometers from the Earth. The Jovian moon is 3,126 km in diameter and its ice crust surface is 18.9 million square kilometers in area, which is about the same size as Africa. Europa's day/night cycle is 3.55 standard Earth days, or 85.2 hours long, which is the same as its orbital period around Jupiter with which it is rotationally locked. Europa receives only 1/25th as much light and heat from the Sun as the Earth and because of its great distance from the Sun its surface temperature at noon is a rather cool $-200°$F!

Attractions

In addition to it being a future tourist attraction, Europa will be a moon of particular interest due to its potential for extraterrestrial life and also as a place for observing Jupiter, the largest gas giant in the Solar System. By the time tourist trips to Europa become a reality it is possible that life will already have been discovered below its icy surface, probably close to hydrothermal vents at the bottom of the ocean. As the surface of Europa is mostly water ice, large expanses with no cracks or craters could be turned into ice rinks, although a Europarized Zamboni will probably be required to smooth the ice! The prospect of ice skating on the surface of Europa will be particularly appealing due to the moon's low gravity, which is only $1.31\,m/s^2$ compared with Earth's $9.8\,m/s^2$, which means that even the laziest skater will be able to perform a triple, quadruple, or even an octuple axel! For those tourists who are more competitive, ice hockey will surely feature as an attraction, although the pressure suits will be a lot more cumbersome than the sports gear worn by terrestrial players. Other sports will no doubt include skiing, a sport which will take full advantage of the pressure ridges and ice fault scarps that are a classic feature of Europan terrain. For those with artistic skills, ice sculptures similar to those displayed in Earth's northern cities during winter festivals will no doubt prove popular. Due to the extreme cold, ice sculptures carved on Europa will never melt, even if they are exposed to full Europa-strength sunlight, which is only 1/25th of the brilliance we are accustomed to in Earth's inner-space location.

Tourist attractions will include a resort-styled Europan Jovian System Observatory that will offer a view of Jupiter "hanging" over the horizon. Although Europa orbits Jupiter at a distance 75% more than the Moon's distance from the Earth, Jupiter is 11 times the diameter of the Earth, so to anyone standing on the surface of Europa it will appear six times as wide as the Earth would from the Moon. This means that Jupiter observers will see the gas giant as a brilliant multi-hued ball in the sky that will fill 550 times more sky than a Full Moon seen from Earth.

Oceanic exploration

One activity that will surely prove popular is the exploration of the Europan oceans, a feat that will be made possible by a descendant of a prototype autonomous underwater vehicle (AUV) that is currently being developed to conduct an unmanned exploration of the Jovian moon. The Deep Phreatic Thermal Explorer (DEPTHX) AUV is capable of submerging to a depth of 1,000 m (3,280 feet) and uses computers linked to sonar information to create 3D images that are overlaid in the computer memory to build a progressive geometrical map, a feature that may prove an attractive memento for future spaceflight participants who eventually venture under

Europa's 10 km thick ocean. In addition to being able to take home some unique extraterrestrial underwater maps, there is the very real possibility that visitors will be witness to Europan life. If a liquid ocean does exist on Europa and volcanic activity from the tidal forces caused by the moon's proximity to Jupiter exists, there may be hydrothermal vents on the ocean floor much like those found on Earth. Here on Earth, diverse lifeforms survive around hydrothermal vents, so there is reason to believe similar organisms may exist on Europa.

The manned equivalent of DEPTHX will be stored on the ice crust surface in a simple modular hangar and will descend into the Europan Ocean using a heated bow cap that will thermally melt the ice. To prevent the ice behind the submarine refreezing, the shaft the submarine creates as it descends downwards will be percolated. As the submarine descends it will reel out a tethered communications cable until the submarine is below the lowest downward protrusions of the ice crust, at which point an antenna will be fixed to the cable, and the cable cut. The submarine will then be free to continue on its intra-oceanic excursion under the Europan ice while maintaining communications with the surface base via sonar to the antenna suspended below the descent shaft. Exactly what Europan tourists may see during these excursions is obviously open to conjecture and speculation. One possible attraction will be witnessing volcanic outgassing from points along the ocean bottom, which will lead to gas building up in the concavities of the underside of the crust. From time to time these gas pockets will exert pressure that will rupture the ice along weak fault lines resulting in the escape of gas into space.

Staying on Europa

Living on Europa will require some engineering ingenuity that will most likely take the form of pressurized chambers located on the underside of the moon's ice crust and Lexan thermopaned geodesic domes and vaults located on the surface. Energy for Europan tourist's habitats will most likely be derived from a process called Ocean Thermal Energy Conversion (OTEC), a means of tapping energy from the heat differences between Europan surface industry waste-heated water reservoirs and cold ocean waters. It is also possible that solar power could be used, assuming improvements in the efficiency and use of concentrating mirrors.

You may be surprised to learn that there is a mission planned to visit Europa, called the Jupiter Icy Moons Orbiter (JIMO), which will launch in 2011. Unfortunately, the mission will be unmanned and there are no plans for any manned missions in the near or even distant future as the third-generation nuclear spacecraft that will be needed to take tourists to Europa are not even on the drawing boards.

EXOPLANETS

An extrasolar planet, or exoplanet, is a planet beyond the Solar System. Although most of the 252 known exoplanets are massive gas giants like Jupiter, there are some that may be worthy of a visit, although there will need to be significant advances in propulsion technology before that happens as one of the nearest possible terrestrial exoplanets is 20 light years from Earth. This planet, named Gliese 581 c, is the third planet of the red dwarf star Gliese 581, and is thought to orbit the habitable zone that surrounds the star. Using conventional rocket technology to travel to these exotic locations beyond the Solar System will not be an option. Perhaps one concept that will be used is the matter–antimatter annihilation propulsion system, which may achieve a specific impulse (a measure of efficiency) of up to 1 million seconds, compared with the Space Shuttle's 455 seconds. Matter–antimatter annihilation, which is the complete conversion of matter into energy, releases the most energy per unit mass of any known reaction in physics. Unfortunately, antimatter is presently the most expensive substance on Earth, at about U.S.$63 trillion per gram, although improvements in equipment used to trap the energy may bring this price down to a more affordable U.S.$5,000 per microgram.

HAZARDS OF EXTREME SPACE TOURISM

Interplanetary debris

Space is full of interplanetary flotsam and jetsam that include chunks of rock, pieces of spacecraft, and dust-sized micrometerioites. Although this hazard is most acute in Earth's atmosphere, where there are as many as 100,000 pieces larger than a centimeter across, the dangers still exist for those embarking on interplanetary ventures to Mars and beyond. For example, the Pioneer spacecraft recorded 55 micrometeoroids impacts between Mars and Jupiter.

Solar flares

The "solar wind" is the stream of electrically charged subatomic particles that flood outward from the Sun at speeds of up to 400 kilometers per second. Occasionally, there are violent outbursts of such particles, known as solar flares, which pose a serious threat to interplanetary passengers, some of which are summarized in Table 8.1.

Health risks of interplanetary flight

Perhaps the greatest challenge for interplanetary spaceflight participants will be dealing with the serious medical problems that occur in microgravity. During your transit to Mars or Europa your bones will shed calcium at an alarming rate, regardless of how much exercise you perform, as the replenishment of calcium cannot keep pace with the rate of loss. The longer your flight time the more brittle your bones will become and the more susceptible you will be to injury once you arrive at your destination. In fact, if you have your heart set on visiting Titan, you may have to come to terms with the fact that your skeleton will be so weak that a return to Earth will be impossible and you may have to consider relocating to the Moon! The

Table 8.1. Risks of traveling to Mars.

	Earth launch	Transit	Mars landing	Mars surface	Mars launch	Transit	Earth return
Source	Van Allen belts	GCR and SPE		GCR and SPE		GCR and SPE	Van Allen belts
Exposure		4–6 mo		18 mo shielded by Mars atmosphere		4–6 mo	
Cumulative exposure	Days		4–6 mo		22–24 mo		26–30 mo
G load	Up to 3G	0 G	3–5 G	0.38 G	2 G	0 G	3–5 G
Notes	Boost phase (8 min)	4–6 mo	Aero-braking, parachute braking (30 s), and powered descent (30 s)	18 mo	Boost phase TEI	4–6 mo	Aero-braking and parachute braking
Cumulative hypo G	0		4–6 mo		22–24 mo		26–30 mo
G transition	1–0 G		0–0.38 G		0.38–0 G		0–1 G

insidious process of osteoporosis is only one of myriad problems you will face, however. The calcium that your body will be shedding every day will circulate in the bloodstream and accumulate elsewhere, eventually manifesting itself as a kidney stone. Despite your three to four-hour daily exercise regime, your muscles will slowly atrophy and your blood cell production within your bone marrow will reduce, leading to anemia and an eventual weakening of the immune system. The heart will also atrophy as it becomes accustomed to a reduced pumping load, a condition which may have serious consequences on return to Earth when the load requirement is increased.

Because of all these problems the first interplanetary tourists will almost certainly be considered as a flying laboratory in their own right and will be used as human "guinea pigs" during and following their flight. In return for a reduction in the cost of their ticket, these tourists will be providing blood samples, wearing sensors, logging their food and drink, storing their waste, and submitting themselves to all kinds of tests in the name of spaceflight medical science.

Living on an interplanetary spacecraft

Spacecraft are noisy places, a fact that will be especially true during the first interplanetary missions, for which equipment will be designed to be functional rather than comfortable. First, there is the brute power that is required for the many electric fans to move air around the cabin. The intakes and outlets that you will see scattered around the cabins will result in drafts chilling your back and drying your mouth. Big noisy boxes called "scrubbers" will recycle old air and remove carbon dioxide using lithium hydroxide filters ensuring that as little oxygen as possible is wasted. Fungal accumulation in the scrubber vents will be another problem that interplanetary tourists will need to take care of on a weekly basis, to say nothing of the inconvenience of the problems associated with the vagaries of the urine-recycling system.

Psychological support

The odors, noise, and unusual color schemes of a spacecraft can each be tolerated for extended periods as long as you have friendly company. Unfortunately, you cannot choose your fellow tourists and long-duration space missions have suffered more than one isolated case of a nervous breakdown. You will therefore have to expect occasional tensions, frustrations, and bickering between crewmembers as well as the inevitable cultural and psychological isolation that may occur as a result of your fellow crewmembers not speaking your language. In the trapped environment of a spacecraft something as simple as food may be enough to cause you to become slightly unhinged. For example, if the smell of a certain nation's cuisine wafting through the cabin annoyed you on the first day of the flight, imagine how annoying it will be on Flight Day 253 or Flight Day 1,056? For those tourists venturing into interplanetary space the first few days of disconnection from Earth will be tolerable, but a few weeks or months of it will inevitably lead to strain.

9

Epilogue

It is difficult to say what is impossible, for yesterday's dream is today's hope and tomorrow's reality

Robert Goddard

Manned spaceflight has captured the public's imagination for over 40 years and it is inevitable that people should ask if and when they too might venture into space. The last 10 years have witnessed an explosion of interest in people traveling to ever more remote areas of the Earth in the search for adventure, and for many of these terrestrial tourists space will be the next logical step. As we have seen in this book, with the advent of commercial launch capabilities directed specifically at tourists, those adventurers are likely to have their opportunity to fly very soon.

The concerted efforts of these private space companies embody the strong will and vision to develop and produce a space fleet capable of not only ferrying tourists to and from orbit but also to keep exploring space. As we have witnessed with the successful flights of SpaceShipOne and Genesis I and II, with the technologies available anything is possible. Although these companies must work with the realities of significant financial constraints, overcome daunting technical challenges, and negotiate minefields of legal bureaucracy, they are now at the threshold of opening the frontier of space to a much greater section of the population. Just as it was difficult for lunar astronauts to describe their feelings when viewing the Earth from the Moon, it is impossible to describe the paradigm shift and ramifications that will occur when the first spaceflight participant views the Earth from Mars. Such a reality will only be possible through the concerted endeavors of visionary companies such as Bigelow Aerospace, Virgin Galactic, and the other privately funded ventures described in this book.

Space has always made an indelible impression on those who have ventured there and it is unlikely that this will ever change. The humbling awareness of the vastness of space will be a profound experience for everyone who has the opportunity to view Earth as a "pale blue dot". The first spaceflight participants will inevitably be those aforementioned adventure tourists, who often perceive themselves as being closer to explorers, usually traveling to places where little or no tourism infrastructure exists, as is the case with space tourism today. Eventually, however, the space tourism industry will evolve and the ranks of spaceflight participants will grow. The exotic travel destination that suborbital flight holds now will inevitably give way to orbital flight, which in turn will give way to lunar excursions and trips to Mars, after which the true pioneers will be those venturing to Titan and Europa. By this time, space tourism will be operating as a fully fledged commercial industry capable of offering you any number of "trips of a lifetime" and truly opening the frontier of space.

Index

Printing: Mercedes-Druck, Berlin
Binding: Stein+Lehmann, Berlin